Oiling the Urban Economy

T0229683

This book presents a critical analysis of the 'resource curse' doctrine and a review of the international evidence on oil and urban development. Its particular focus is the role of oil in property development and rights in West Africa's new oil metropolis – Sekondi-Takoradi, Ghana.

It seeks answers to the following questions: What changes to property rights are oil prospecting, explorations, and production introducing? How do the effects vary across different social classes and sectors of the economy? To what extent are local and national institutions able to shape, restrain, and constrain transnational oil-related capital accumulation and how it affects property in land, property in housing (residential, leisure, and commercial), and labour? How do these processes impact on Sekondi-Takoradi and connect with the entire urban system in Ghana?

This book shows how institutions of varying degrees of power interact to govern land, housing, and labour in the city, and analyses how efficient, sustainable, and equitable are the outcomes of these interactions. It is a comprehensive account of the tensions and contradictions in the urban economy, society, and environment in the booming oil city. It should be of interest to urban economists, development economists, real estate economists, Africanists, and geographers.

Franklin Obeng-Odoom is currently the Chancellor's Postdoctoral Research Fellow at the School of Built Environment at the University of Technology, Sydney, Australia.

Routledge studies in international real estate

The Routledge Studies in International Real Estate series presents a forum for the presentation of academic research into international real estate issues. Books in the series are broad in their conceptual scope and reflect an inter-disciplinary approach to Real Estate as an academic discipline.

Oiling the Urban Economy
Land, labour, capital, and the state in Sekondi-Takoradi, Ghana
Franklin Obeng-Odoom

Real Estate, Construction and Economic Development in Emerging Market Economies
Raymond T. Abdulai, Franklin Obeng-Odoom, Edward Ochieng and Vida Maliene

This is a penetrating analysis of how a fortuitous resource endowment impacts on a developing economy. It shows that having oil resources creates major tensions – social, environmental and political – as well as potential economic benefits. Obeng-Odoom's book carefully reviews the evidence, drawing on useful currents of political economic analysis. It also proposes policies that could yield better outcomes. The book is warmly recommended to all who want to learn the lessons from the Ghanaian experience.

Frank Stilwell
Professor Emeritus in Political Economy
The University of Sydney

A well-written book that analyses oil cities and their potential development impacts from a heterodox perspective. Obeng-Odoom focuses on the small and unknown twin city of Sekondi-Takoradi to elicit how geologic resources, such as oil and gas, contribute the fiscal resources for social and economic change in new oil cities. This central argument is couched within the context of how human agency, in the form of unions, ensures a fairer distribution of the outcomes of development. This book is a must-read for intellectuals interested in how urban and rural development in the global periphery is grounded in alternative political-economic and ecological frameworks.

Ian E.A. Yeboah
Professor of Geography
Miami University, Oxford, Ohio, USA

Oiling the Urban Economy

Land, labour, capital, and the state in
Sekondi-Takoradi, Ghana

Franklin Obeng-Odoom

LONDON AND NEW YORK

First published 2014
by Routledge
2 Park Square, Milton Park, Abingdon, Oxon OX14 4RN

and by Routledge
711 Third Avenue, New York, NY 10017

First issued in paperback 2018

Routledge is an imprint of the Taylor & Francis Group, an informa business

British Library Cataloguing in Publication Data
A catalogue record for this book is available from the British Library

Library of Congress Cataloging-in-Publication Data
Obeng-Odoom, Franklin, author.
  Oiling the urban economy: land, labour, capital, and the state in
  Sekondi-Takoradi, Ghana/Franklin Obeng-Odoom.
  pages cm. – (Routledge studies in international real estate; 1)
  1. Petroleum industry and trade–Economic aspects–Ghana. 2. Petroleum
  industry and trade–Political aspects–Ghana. 3. Resource curse–Ghana.
  4. Land use, Urban–Ghana–Sekondi. 5. Land use, Urban–Ghana–
  Takoradi. 6. Sekondi (Ghana)–Economic conditions. 7. Takoradi
  (Ghana)–Economic conditions. I. Title. II. Series: Routledge studies in
  international real estate; 1.
  HD9577.G42O24 2014
  338.2728209667-dc23
                                                              2013046466

ISBN 13: 978-1-138-62607-2 (pbk)
ISBN 13: 978-0-415-74409-6 (hbk)

Typeset in Times New Roman
by Wearset Ltd, Boldon, Tyne and Wear

Contents

Illustrations

Figures

Photographs

Tables

Preface

The discovery of oil in Africa is typically accompanied by alarmist reports in the press, especially the Western media, asserting that oil will either suddenly make its possessor corrupt, dependent, or anomic. These reports impede our understanding of expectations and the ramifications of oil exploration, drilling, and exportation for urban economic development in Africa. They tend to ignore the complex processes involved in the creation of oil rent, the nexus between potential oil rents and actual oil rents; oil rents (in all their shades) and land rent, oil rents and housing prices, and oil rents and changing livelihoods. Also, they overlook how these connections are shaped by colonial, neocolonial, and neoliberal forces and structures, and what these different relationships mean for contemporary dynamics of livelihoods, their adaptation, and adoption, urban and housing form. Further, they shed little light about urban and regional expectations, how they are constructed, transformed, lived, and met, and how they, in turn, relate to land, labour, capital, and the state. It is these nuances that this book explores in the context of Sekondi-Takoradi, a West African oil frontier that the British once considered to be Ghana's Eldorado.

I first encountered Sekondi-Takoradi when I enrolled at Bishop Essuah Memorial Catholic Primary School in Roman Gate, Takoradi in 1990. I lived in the city until my grandfather, a post-colonial judge, retired from the Judiciary Service in 1994 and relocated to Cape Coast which at the time was much quieter than Sekondi-Takoradi. I schooled in Cape Coast until 1997 and then left for Kumasi, the second largest city in Ghana, to continue my education, including my undergraduate studies in land economy in 2005. It was when I was in England, a year later, to pursue a master's programme at the University of London (UCL) between 2006 and 2007, that Ghana discovered oil off the shores of Sekondi-Takoradi. One of my teachers in England, an urban planner with interest in Ghana, expressed worries about the discovery, pointing out the experiences of oil in other African countries. I started thinking about oil very carefully but the interest in this resource all around me was couched in 'resource curse' or 'oil bonanza' terms. Even when I started a PhD in political economy in 2008, a year after the discovery of oil, there was no study on the urban expectations and experiences of oil, their gendered and generational aspects, colonial and colonising gaze, or their different and differentiated dynamics. So, alongside writing my doctoral thesis, for which oil was not my central concern, I started vigorously studying oil and, because of my experiences and interest in land economy, urban economic development, and political economy, focused on the urban question within the oil experience – both temporally and spatially. This emphasis brought me back to my 'roots', Sekondi-Takoradi now globally regarded – albeit poorly studied – as Ghana's 'oil city'.

Writing of the book started in 2008/2009 and has gone through three phases since then. The first phase (2008/2009 to mid 2012) entailed desk review, analysis of published studies on oil, and telephone/electronic communication with key stakeholders involved in the management of the oil city or general policy making in the oil city. The second phase (end of 2012 to mid March 2013) commenced when research funding was sought and obtained to facilitate primary, fieldwork research. Activities in this second phase included living and researching in Sekondi-Takoradi. The final phase (end of March 2013 to February 2014) entailed further analysis and write up of the evidence I collected and that compiled by others.

Writing about Sekondi-Takoradi has been extremely challenging. The 'capital city bias' in urban research plagues Ghana too, so most of the scholarship on urban Ghana is on Accra and inspiration for Sekondi-Takoradi was not easy to find. In turn, I had to draw on the expertise and help of many scholars and practitioners to complete this book. I acknowledge the monumental help of Mr. Raphael Edem Fiave of Sekondi-Takoradi Metropolitan Assembly (STMA) for his support and research assistance before, during, and after my fieldwork in Sekondi-Takoradi. Special thanks also to all my interviewees, especially at the Esikado Palace, the Ghana Tourism Authority, the Rent Control Department, and the Ghana Revenue Authority. Special thanks to the workers at the Ghana Railway Company, Takoradi, who spent a lot of time with me sharing fond memories of the Railway and reflecting on its current plight. I appreciate the help of the land economists and surveyors at Asenta Properties Ltd, Takoradi, especially the Head of Valuation and Estates Services, Surveyor Frank Opoku. The encouragement and support of Joe Prempeh-Bentil, based at the University of Mississipi, USA, are greatly appreciated as is the detailed, critical, and very helpful feedback I received on various chapters from Dr Robbie Peters, Dr Bob-Milliar, and Assoc. Prof. Stuart Rosewarne, Dr Annie Herro, Dr Ransford Gyampo, Mr David Primrose, Mr Thomas Allen, and Ms Kim Neverson. Prof. Ian Yeboah deserves special mention for reading and editing the entire manuscript. I benefitted greatly from his experience and insights as a senior urban scholar, specialising in Ghana.

I must also thank the members of staff of the Sekondi-Takoradi Workers' College, especially Mr Kondja, Mr Cudjoe, and Mr Larweh who made my stay at the Workers' College enjoyable. Prof. Spike Boydell, my senior colleague at University of Technology, Sydney, deserves special appreciation for his unwavering support and interest in my career development and for the great lengths he went to provide me mentorship during my research fellowship which led to the writing of this book, and University of Technology, Sydney, for offering me research funding for the study.

Many others deserve appreciation. I wish to single out and acknowledge the help of my friend, teacher, and mentor, Prof. Emeritus Frank Stilwell, who carefully went through a draft of the monograph and, in his usual way, gave me extensive and helpful feedback. I am also grateful to my editor, Mr Ed Needle for his continuing encouragement. I thank the publisher's reviewers whose comments have greatly helped to strengthen this book and thank reviewers of the following journals: *Development and Society, Cities, Urbani Izziv, City, Culture and Society, Local Economy, African Review of Economics and Finance, The Extractive Industries and Society,* and *Local Environment* from whom I have learnt a great deal in my research on 'oil cities' and whose comments have helped to strengthen aspects of this book. I acknowledge the kindness, encouragement, and unwavering support of my family, especially my grandmother, Rose; my mother, Emily; and my sisters, Clarissa and Ophelia. Finally, I am deeply grateful to my beloved wife, Dr Hae Seong Jang, Korean sociologist, teacher, and researcher, for her care, patience, and advice without which this book would not have made it to print.

Franklin Obeng-Odoom, PhD

The Chancellor's Postdoctoral Research Fellow and

World Social Science Fellow

School of Built Environment

University of Technology, Sydney, Australia

Part I
The economics of 'black gold'

Photo 1 Religious poster about transforming a curse into a blessing.

Photo 2 A closed shop in Cape Coast – echoing the dichotomy of blessing and curse.

1 Africa's oil wealth and crude interpretations of its ramifications

Introduction

Africa is one of the richest continents in the world. Alone, it contributes a massive 11 per cent of the global oil production and possesses a further 8 per cent of the world's total oil reserves (African Development Bank *et al.*, 2013, p. 37). As of 2007, Nigeria, Angola, Cameroon, Chad, Congo-Brazzaville, Equatorial Guinea, and Gabon were collectively producing an estimated 5,120 million barrels of oil per day. Also, new oil fields are being developed (Basedau and Mehler, 2005, African Development Bank and African Union, 2009; African Development Bank *et al.*; 2013). Africa can, therefore, be labelled as a wealthy continent in terms of its natural resource endowment.

While the quantity of its oil reserves may not be analogous to that of the Middle East, the upward trend of oil discoveries makes the continent a focal point of attention for traditional Western powers and emerging global players in Asia, particularly China (Basedau and Mehler, 2005; Mohan, 2013). In the last 20 years, new discoveries of oil reserves in Africa have shot up by 25 per cent whereas those of gas have increased by 100 per cent (African Development Bank (AfDB) and the African Union (AU), 2009). The leading producers of oil on the continent are Libya, Angola, Nigeria, Algeria, and South Sudan, while Egypt, Libya, Angola, and Nigeria top the list of gas producers. These countries collectively produce more than 90 per cent of the oil and gas reserves in Africa (African Development Bank and the African Union, 2009).

A well-known chorus in social science analysis is that, in spite of Africa's oil wealth, most oil and gas endowed countries have not been able to use the revenues from crude extraction to enhance human development for their people. Instead, economists typically argue that the revenues from oil make regimes and governments repressive. In Libya and Chad, a substantial part of the oil revenues go to military spending at the expense of guaranteeing civil rights and liberties. Cameroon is another case in point, where some analysts argue that oil has made it difficult for local governments to improve participation (Oyono *et al.*, 2006). Angola has had similar experiences. According to Nicholas Shaxson (2009), its 2009 budget shows significant returns from oil, totalling about $42 billion, bigger than all the overseas development assistance the OECD countries give to the entire African continent. In fact, oil production in Angola has kept rising: from 1.2 million barrels per day in 2005 to 1.6 million barrels per day in 2007 and an estimated 2.0 million barrels per day in 2009. Yet, Angola struggles with widespread conflict, violence, and corruption, while social services are limited (Same, 2009, p. 7). In Chad, oil from the Chad–Cameroon Pipeline alone stood at 140,000 barrels per day in 2004 and, by the first half of 2005, it had increased to 180,000 barrels per day. Again, research shows that, in terms of social effects at the household level, the oil seems to have had limited impact. A substantial part of the revenue goes into military spending, leaving public spending on health low and 80 per cent of the population in poverty (Pegg, 2006). These experiences are not mere correlations; orthodox analysts argue: oil causes them all.

Indeed, orthodox economists lump these problems together and refer to them as 'resource curse'. As a concept in economics, the notion of a 'resource curse' connotes the inverse relationship between resource wealth and the health of an economy (Corden and Neary, 1982). Hence, the resource curse is sometimes called 'the paradox of plenty' (Auty, 1993). The curse can plague all continents. However, Africa has become the focal point of considerable research and interest (Obi, 2009), owing to what is widely perceived to be a particularly endemic curse in that region (Lesourne and Ramsay, 2009), feeding into widespread allegations of neopatrimonialism.

According to Sala-i-Martin and Subramanian (2012), there are three ways to explain the inverse relationship between resource boom and the 'paradox of plenty'. First, resource wealth increases the chances of rent-seeking behaviour which may lead to corruption and its concomitant negative effects on the economy. Second, dependence on oil makes a country vulnerable to global fluctuations in oil prices. Third, investing in oil may drive out investment in other sectors or increase the value of the local currency, making local exports uncompetitive (Dutch disease effects). A complementary reason can be found in the Prebish–Singer model, which suggests that the prices of primary commodities such as oil are volatile and typically tend to decline over time compared to manufactures. Another suggested reason for the resource curse is that wealth obtained without sweat leads to indolence (Sachs and Warner, 1995, 2001; Collier, 2006, 2008, 2009; Hammond, 2011; Sala-i-Martin and Subramanian, 2012; Yates, 2012). All these perspectives view oil as a curse.

The oil curse has been particularly pervasive in Nigeria. One of the wealthiest countries in Africa, it is a major source of oil and gas in the world, producing about 2.46 million barrels of oil on a daily basis, about 22 million tonnes of gas per year (Obi, 2009). Between 1970 and 1999, $231 billion in oil revenues accrued to Nigeria (Ross, 2003). Most of this oil comes from the Niger Delta area (Obi, 1997), home to about 33 billion barrels of crude oil reserves. Despite all this wealth, social conditions in the area remain poor (Aaron, 2005; Obi, 2007; Egiegba, 2013). An estimated 70 per cent of the people in the Niger Delta area have no clean water, passable roads, electricity supply, and low medical supplies (Idemudia, 2007, p. 3). In the oil-rich states of Delta and Bayelsa, the patient–doctor ratio is 150,000 to one (Watts, 2006). Between 1970 and 1999, revenue from oil rose but per capita income fell (Ross, 2003). An estimated 80 per cent of Nigeria's oil accrues to only 1 per cent of the population, while the remaining 99 per cent of the population struggles to share only 20 per cent of the oil revenue (Obi, 2009, p. 123). There are also severe environmental problems in the form of pollution and land degradation in the Niger Delta (Obi, 1997; Aaron, 2005), which is also a source of recurrent violent conflict. Several foreign oil workers have been captured. In 2007, nine Chinese workers were abducted and a car bomb detonated (Obi, 2008, p. 418). In 2004, an estimated 50 people died from oil violence per day (Arowosegbe, 2009, p. 588).

While orthodox economists (e.g. Collier, 2009) view these struggles as result-ing mostly from criminal activities such as vandalism and sabotage, political economists regard them as deriving from structural reasons such as social depri-vation (Aaron, 2005; Shaxson, 2007; Egiegba, 2013), hierarchical clientelism (Ifeka, 2004), and a symbiotic relationship with the state, local elites, and trans-national corporations that inhibits the effectiveness of institutions advocating transparency in the oil sector (see also Watts, 2004). To these concerns, Obi (2009) adds ineffective land ownership. He argues that central control of locally owned land sees elites from the majority ethnic groups control the oil produced from the minority states. These factors are intensified by local (e.g. elites in the Niger Delta leadership), national (e.g. officers of the state and the military), and global (e.g. oil multinational corporations) forces. The local elite forces have two faces, one for the people and the other for their self-gain. The federal state perceives the struggles as attempts by criminals to bring the Nigerian economy to a halt.

At the root of the problem are questions of self-determination and resource control (Obi, 2008, p. 430) and questions of neglect, particularly of the youth in the Niger Delta (Arowosegbe, 2009) but also of a general feeling that the wealth of the area goes to benefit other regions (Aaron, 2005). Revenue allocation for the development of the area continuously fell between 1953 and 1989. From 100 percent in 1953, revenue allocation fell precipitously as follows: 50 percent (1960), 45 per cent (1970), 20 percent (1975), 2 percent (1982), and 1.5 percent (1989). It rose to 3 per cent in 1993 (Aaron, 2005), and to 13 per cent of oil revenue in 1995 but, by 1999, only 1 per cent of oil revenue was paid to the Niger Delta area as derivation (Imobighe, 2011). Within these broad time frames, there were times, for example, in 1977 and 1981, when the Abayode Commission and the Okigbo Commission respectively removed the derivation principle (Imobighe, 2011). As of 2005, it was 13 per cent (Aaron, 2005) and this share remains to date; however, it is widely believed that the effects of deri-vation are yet to be felt in the area (Egiegba, 2013).

There seems to be considerable interest in improving the situation in the Niger Delta area. As of late 2009, the federal government had relinquished 10 per cent of its stake in oil to the local Niger Delta communities, complementing another gesture known as the Amnesty Programme. Amnesty entails disarma-ment, rehabilitation of former militants, and infrastructural development in the

Niger Delta area. While it is a major improvement over previous policies of anti-criminalisation, the programme is only narrowly focused on militants, not other locals who suffer similar fates as the militants yet did not take up arms (Oluduro and Oluduro, 2012; Ugor, 2013). These nuances are not considered in the universally teleological, if not theological, reading of resource experiences in Africa.

Indeed, even in cases where oil and gas production generate substantial economic benefits, there are notable social and environmental problems. Such was evidently the case for Oloma, a fishing community on the Island of Bonny in the Niger Delta area. Oil seemed to have improved infrastructure, education, and banking in the area. Overall, however, researchers found that the life of the ordinary people was adversely affected. Fishing became less lucrative as oil installations and trucks polluted the waterways and the sea. Spillage was a key concern, as was noise. Land filling for oil operations, canalisation for oil machinery, erosion, and gas flaring all harmed the environment. Drinking water was contaminated, and people developed skin rashes and respiratory disorders because of oil and gas production (Fentiman, 1996). A recent special issue of *American Review of Political Economy* on 'Africa-based Oil Exporting Companies: The Case of Nigeria' (Snapps, 2011) and an edited book entitled *Natural Resources, Conflict, and Sustainable Development: Lessons from the Niger Delta* (Okechukwu et al., 2012) affirm that, to date, the social and ecological footprint of the oil industry has been devastating, while prodigious rents have accrued to economically and politically powerful groups in the country, including top politicians.

Another case that problematises the resource curse doctrine is that of Botswana. Between 1966 and 1999, its economy grew by an average of 9 per cent per annum, making it the fastest growing economy in the world at the time. Also, relative to other resource-rich countries in Africa, it has succeeded in investing resource revenues into social development, particularly education and health (Pegg, 2006). Before the transformation of its economy and society in the 1970s, through the discovery of diamonds, Botswana was regarded as a 'poor country dependent on livestock, remittances and foreign aid' (Poteete, 2007). Gaborone, its capital city, has recently been described as 'extraordinary in African terms ... lacking in mass poverty, extensive squatter settlements or civil strife' (Kent and Ikgopoleng, 2011, p. 478). Indeed, the 2010 *Human Development Report* (UNDP, 2010, p. 30) describes Botswana as one of the few countries in Africa to have made 'substantial progress in human development'. Botswana demonstrates that the experiences of resource wealth in Africa differ widely and that categorical claims based on the resource curse doctrine must be tempered.

In the case of Africa's oil cities,[1] the resource curse doctrine is particularly problematic because it was formulated without taking into account the dynamics of space, urbanity, and urbanism. We know relatively little about oil cities in Africa except to guess what their experiences might be on the basis of the resource curse idea, and its overemphasis on national factors (Ayelazuno, 2013).

So, there is the need for a substantive study of Africa's oil cities in their own terms to understand in what ways or to what extent they confirm, confound, or conform to the stereotypes.

Existing studies are inadequate for understanding oil cities in Africa.[2] There is a small literature on mining and cities, consisting mainly of James Ferguson's work on urban settlements on the Zambian copper belt (e.g. Ferguson, 1990a, 1990b), Brycesson's research on Katoro, a trading town adjacent to the Tanzanian mining city of Nyarugusu (Brycesson, 2011), Walsh's work on gold mining and trading urban centres in Madagascar (Walsh, 2012), Kent and Ikgopoleng's (2011) analysis of Gaborone's experiences with diamond mining, and Gough and Yankson's (2012) work on gold mining in Obuasi, Tarkwa, and Kui in Ghana. Other case studies, including of diamond mining, are available in the 2012 special issue of *Journal of Contemporary African Studies* (Bryceson and MacKinnon, 2012). These studies do not consider oil which differs significantly as a natural resource from other minerals (e.g. gold) and therefore they cannot be applied to oil cities. Crucial to the global economy (Huber, 2013), oil drilling is also so capital intensive that it is rare for individual miners to attempt it, let alone succeed in doing it, unlike in the case of gold mining (Kenamore, 1983; Banchirigah, 2008).

It is important to focus directly on oil cities themselves, given that Africa is becoming increasingly urbanised, with more people moving to oil cities. Oil cities such as Alexandria have been making massive national and regional contributions (UN-HABITAT, 2008b, 2010). This book examines the case of Sekondi-Takoradi in Ghana, West Africa. Sekondi-Takoradi is a good example not only because it is a new oil city but also because, as is typical of Africa's so-called 'secondary cities', little is known about it (Lawrie *et al.*, 2011; Brycesson, 2011; Swing *et al.*, 2012). Indeed, the first and only regional human development report in Ghana, the *Western Region Human Development Report 2013* (UNDP, 2013), made just 21 brief references to it (nine to Sekondi and 12 to Takoradi), mostly in the appendix, although Sekondi-Takoradi is the region's most prominent city. The reason for this neglect, according to the report, is that we do not, as yet, understand this oil city and hence it called for more research to fill this major lacuna (UNDP, 2013, p. xv). Another recent study, published in *Africa Today* (Andrews, 2013, p. 70), has highlighted the need for studies on urban economic development and oil. So, a focus on Sekondi-Takoradi is crucially important.

Once described by a colonial Correspondent (1943, p. 38) as 'a small collection of dirty reed and thatch huts where the beach ended and the bush began', West Africa's newest oil zone has captured international attention. The Government of Ghana is optimistic about its future, as are corporate interests and workers in the oil sector (McCaskie, 2008). The expression of excitement on the oil rig itself is worthy of note; chanted by the oil rig workers, it went, 'We made it for Ghana! We made it for Ghana!' (*Offshore*, 2011, p. 13). But, alongside the optimism is widespread pessimism. Some chiefs, the custodians of most of the land in the Sekondi-Takoradi metropolis, are concerned about whether compensation will be paid for traditional land and fishing rights that risk being destroyed as a result of the resource development process (Boohene and Peprah, 2011). The chiefs themselves are the focus of suspicious gaze as are most of their theatrical public performances intended to evoke feelings of pristine or pure traditions or cultural ethos. Most of the youth fear that the benefits of oil will not be felt locally. Civil society groups and others believe that there are issues of accountability yet unsolved while legal experts believe new laws and regulations are needed without which the future of the country will not be bright (Friedrich-Ebert-Stiftung, 2011).

The twin city of Sekondi-Takoradi itself is rapidly becoming host to speculators, migrants, and investors. Richard Grant (2012), a leading scholar on urban economic development in Ghana, has observed that oil will lead to a reconfiguration of the metropolis, reintegration into the national urban system, and further property rights ramifications. These are empirical matters that require substantial and substantive investigations. This book seeks to do so by providing the results of substantive investigation into the concerns and contradictions emerging from what has been happening in and around Sekondi-Takoradi.

Aims and objectives

This book does five things. First, it problematises the resource curse doctrine at the macro and urban levels. Second, it demonstrates the class nature of the expectations and ramifications of oil at the urban level. Third, this book shows how colonial experiences can shape the nature of present day institutions, investigating how those dynamics constrain the experiences of oil cities. Fourth, it analyses the various ways in which neoliberalism over the long haul has encountered urban institutions within oil-rich economies, focusing on the relationship between the natural and the built environment. Finally, it considers and appraises existing and proposed social policies that have been developed to deal with the impacts of oil extraction in the region.

The book is written from a particular perspective of urban political economy. Analytically, it is at the intersection of economic sociology, economic anthropology, and economic geography, attempting to understand oil cities as conditioned by underlying economic interests and processes; and how they have evolved within and between different classes with differential power and rights. Simultaneously, attention is given to the intervening role of institutions, and how they, in turn, have changed and are likely to transform under the prevailing global capitalist order. Like the idea of 'mineralised urbanisation' as a 'new wave of urban growth and settlement experimentation is coalescing' (Bryceson and MacKinnon, 2012, p. 514), the ramifications of resource exploration, drilling, and exportation for urban economic development are vigorously investigated in this book. But, so too is the nexus between potential oil rents and actual oil rents; oil rents (in all their shades) and land rent, oil rents and housing prices, oil rents and changing livelihoods. These connections are shaped by colonial, neocolonial, and neoliberal forces and structures, so we need to know what these different relationships mean for contemporary dynamics of livelihoods, their adaptation, and adoption, urban and housing form. In addition, the book examines urban and regional expectations, how they are constructed, transformed, lived, and met, and how they, in turn, relate to land, labour, capital, and the state. It is the attempt to show the connections between these aspects that defines the contribution that this book makes.

This is a departure, therefore, from existing studies of the global oil industry which have tended to focus on the notion of the 'resource curse', as posited by Corden and Neary (1982) and developed by Auty (1993), Sachs and Warner (1995) and others (e.g. Collier, 2009; Kolstad and Wiig, 2012). This book's focus on the urban dimensions of the political economy of oil distinguishes it from other recent research (e.g. Carmignani and Chowdhury, 2011). Indeed, natural resource economics, generally, tends to overlook the urban dimension of oil, as suggested by Robert M. Solow, Nobel Prize winner in Economic Science (see Solow, 2008).

The book presents a critical analysis of the resource curse doctrine but is also critical of other generalisation theses, such as the 'staples' view that sees oil production as entirely good for economic development where countries have hydrocarbons in abundance. Instead, the book presents an original synthesis. In doing so, it directly addresses questions such as: In what ways did the twin city come into existence? What changes to property rights are oil prospecting, exploration, and production driving in the twenty-first century? How are these expected to change the society, environment, economy, and polity? Will the effects vary across different social classes and sectors? To what extent are local and national institutions able to shape, restrain, and constrain (mostly) transnational oil-related accumulation and its effects on property in land, housing (residential, leisure, and commercial), and labour? How do these processes connect with the entire urban system in Ghana? The different scales of analysis range from the mineral sites or settlements themselves, through the wider Western Region, to national and international considerations (Bryceson and MacKinnon, 2012). The answers to the questions are also linked to the growing African and global literature on boomtowns and the social dysfunction or dislocation that oil prospecting, exploration and development creates in local economies (Lawrie *et al.*, 2011; Swing *et al.*, 2012).

The analysis presented here seeks to be significant and innovative in five ways. First, it uses urban analysis. This is significant because, although the oil industry has macroeconomic and global effects, its impacts on particular urban centres where it is located are peculiar, complex, and unlike the national effects or international ramifications. Second, it changes the focus from the global literature on 'peak oil/cities after oil' to 'cities before and during oil'. Peak/post oil research dominates the relatively few papers on urbanism and oil exploration (e.g. Atkinson, 2012) but, while important, it ignores current complexities that have ramifications for the future (peak and post-peak period). Third, the book innovatively draws on ideas from Henry George, David Harvey, Hossein Mahdavy, and Chibuzo Nwoke in developing a spatio-temporal analyses of oil.

This explicitly theorised political economic approach also differentiates it from previous studies. The analysis entails considering how the rewards for factors of production arise, how they are distributed, and how those prior processes affect economic development. Fourth, this book disaggregates the impacts of oil along class lines. This distributive dimension of the analysis is important because the mode and relations of production determine substantially how people experience oil development when driven by capitalist interests. Finally, the book analyses a frontier urban settlement through the oil lens, rather than looking at a capital city. This helps to correct the 'capital city bias' that has existed in African urban studies.

Beyond these distinctive characteristics, the book has the general characteristics of political economic analysis. It emphasises evolution in historical time, how institutions of varying degrees of power interact to shape how land, housing, and labour are used, and how efficient, sustainable, and equitable are the outcomes. There is an emphasis on the tensions and contradictions in the main sectors of the urban economy, society, and environment. An analysis of these concerns, relating to the generation and distribution of oil and land revenues between urban, national, international, and supranational bodies is crucial to understanding the political economy of oil and sustainable urban development in Africa. *Oiling the Urban Economy: Land, labour, capital, and the state in Sekondi-Takoradi, Ghana* is the first book on Africa's oil resources cast in an urban setting.

Sekondi-Takoradi is undergoing a great transformation. The twin city is likely to become a key destination for many migrants in Africa working in different sectors, both up and downstream from oil extraction. While this transformation has many positive ramifications, such as increasing the rate of house building, creating jobs for people, and opening up possibilities of increasing revenue for some state institutions, these opportunities are unevenly distributed across different classes and social groups. Chiefs, bourgeois natives, landlords, house owners, and expatriates have scooped the lion's share of the oil windfall, the land revenues windfall, and the investment opportunities that have accompanied the oil industry. Weaker groups – women, fishers, peasant farmers, poor tenants – have been made worse off and turned into wage labour with poor labour rights. The oil companies extract more rents relative to the state, local, urban, or national. While the expectations for this great transformation have been hyped locally, regionally, nationally, and internationally, it turns out that most are not met and are probably unsustainable.

There are prospects to overturn the social and economic problems bedevilling Sekondi-Takoradi, but these avenues for change are contingent rather than assured. Further, there are grave systemic issues that require urgent focus, but the direction of public policy is clearly not in that aspect. Political economic processes of so-called 'economic recovery' have led to the downsizing of the state, while the contemporary neoliberal outsourcing the responsibilities of the state has systematically nibbled away at the state's capacity for reasoned and efficient pro-poor interventions. The urban authorities, once regarded as 'socialist minded', are now singing the same tune from the neoliberal hymn book – modernisation, outsourcing and privatisation.

In turn, property relations are changing rapidly in the city to catapult the rich and landed elite to new echelons of influence. Worker activism proved invaluable in circumstances like these, especially in the early days of Sekondi-Takoradi when the Railway Workers' Union recurrently stood up for the rights of the exploited and expropriated. Yet, the opportunities of oil have by-passed the railway workers who, in the past, combined strikes, positive defiance, and demonstrations to reclaim the city for workers and to make significant progress in subordinating foreign to national interest, and privilege to common people's interest. While several social and civil society groups currently exist, including the Oil Rig Workers' Union, their size and their location offshore and in another city do not give positive signals as to what extent the workers are united to demand their fair share relative to foreign interests, the interests of the landed elite, and the interest of the few black African moneyed groups. Although the media and civil society groups can keep attention on the massive excesses of the oil and gas industry, they are currently caught in the fever of 'resource curse' and have overlooked the dynamics of land, labour, capital, and the state in Sekondi-Takoradi itself.

Chapters

The book is divided into ten chapters that theoretically and examine the ramifications of oil exploration and production for urban land, labour, capital, and the state. They also examine the opportunity cost of the current use of oil revenues. The chapters cluster in three parts, made up of a critique of conventional approaches to the economics of oil (Part I); early history of the oil twin city, changes in expectations, livelihoods, property relations, and experiences of different social groups (Part II), and some posited policy considerations for a more socially just city (Part III).

This introduction chapter has set the study in the broader context and literature of the political economy of oil. It has emphasised the gaps that exist in our knowledge, and described why and how this book seeks to fill those lacunae. Further, it has introduced the key research questions and describes the structure of the book.

Chapter 2 discusses various frameworks for studying the political economy of natural resources, such as the resource curse doctrine, the factor endowment hypothesis and the open access exploitation thesis. Emphasising the serious inherent drawbacks of these frameworks, it develops a plural property rights framework. This provides critical perspective, drawing from works by Henry George, David Harvey, Hossein Mahdahvy, and Chibuzo Nwoke and linking urbanism with natural resources and economic development. The chapter also describes how that framework is applied to the present study by showing the nature of the data that are used in the book and how they were collected and analysed. The central argument is that, while there are a number of theories that try to explain and predict the relationship between resource abundance and economic development, they do not capture the complexities of oil at the urban level. The neoclassical economics notion of 'resource curse', the political science construct of 'rentier state', and the sociological idea of 'anomie' and 'social dysfunction' pay in sufficient attention to the role of oil in urban society, economy, and environment. This chapter, therefore, argues the case for using a plural political economic framework built around the ideas of George, Harvey, Mahdavy, and Nwoke as a more satisfactory 'way of seeing' how the discovery, exploration, and production of oil affects urban economic development.

Chapter 3 begins the analysis and evaluation of the expectations and ramifications of the nascent oil and gas industry for urban development by providing a critical account of the oil industry in Ghana, focusing on its origins, players and current characteristics, and problematising the resource curse doctrine. The chapter questions the binary representation of the political economy of oil into 'blessings' and 'curses'. It shows that curses and blessings co-exist, intermingle, and impact differently on different social groups. Further, there are many impacts in between the two which are neither curses nor blessings. This evidence suggests that, beyond euphoric reactions and propositions that strike a determinist relationship between resource boom and curse, there is room for practical steps to remedy specific weaknesses in existing public policy. The chapter closes by pointing to the need for an urban level study of the resource curse doctrine.

Chapter 4 introduces Sekondi-Takoradi from a historical perspective in order to provide a context for understanding and appreciating the account of the ramifications of oil post 2007. Therefore, it lays the grounds for an analysis of continuity and change in the oil city. It shows the trajectory of Sekondi-Takoradi, analysing how it emerged, how it rose to fame, and why, until recently, it commanded little national and international attention as a twin city. More fundamentally, it shows how seeds of neocolonialism were sowed in the city, to help explain how these political economic experiences have undermined the quality of public institutions today. It is an account that shows the rapid change from a twin city which had a socialist character to a modernist, capitalist city with vestiges of the past.

Following that, Chapter 5 examines the changing property relations in Sekondi-Takoradi during the period of oil exploration, production, and development. It considers expectations, the changing urban livelihoods on- and offshore and the changing land relations in the city, while analysing ways in which land prices, land access, and land distribution have transformed and impacted differently on diverse socio-economic groups since the discovery of oil. Further, it connects changes in the land and real estate sectors to employment, and examines the implications of the activities therein on labour size, conditions, and characteristics. Again, it shows increasing activity in the residential and leisure real estate market, and tracks the emergence of new real estate agency firms and forms, the intensification of informal estate agency activities, and the development of new alliances in the estate agency sector of the urban economy. Finally, it looks at how private capital is taking advantage of the boom to introduce credit to fund transport investment. It argues that the transformation in the economy has benefitted only a few: the propertied class, landowners, and rich and influential people from bigger cities and other countries. The rewards from the oil industry show that expectations are exaggerated and claims of benefit were over-hyped or inflated. The oil sector has bypassed a majority of people in the city, especially women, indigenes, and poor tenants. While those indigenes who have benefitted from direct employment have improved their economic conditions, they face uncertain career paths, live in fear of being fired, and are consigned to the lower rungs of the job ladder whereas, for most urban youth who espy or aspire for jobs, those desires remain merely aspirational.

Chapter 6 zooms in on the uneven socio-economic effects on the oil opportunities. It focuses on livelihoods in particular sectors, namely fishing and farming, and investigates the expectations, the reality, and the experiences of fishers and farmers (two of the dominant work groups in the city). The chapter reveals that, while there seems to be no clear evidence of massive pollution and hence dispossession of the populations living off the land and the sea in the oil 'zone', there are tell-tale signs of difficult days ahead, related to enclosures and expropriation. Most fishers have experienced a dramatic decline in fishing harvest and expect a worsening of their current plight, while many peasant farmers have lost their land. Similarly, there seems to be actual reduction in livelihoods of fishers and farmers. Institutional strengthening may help resolve the past concerns and impending tensions, but such reform is incapable of addressing systemic problems of growing imbalance in the power relations among the players and pawns in the oil industry.

Chapter 7 examines the prospects for ameliorating the afore-listed challenges through urban and national state planning solutions, focusing on the use of two instruments, namely compensation and betterment. It is important to examine the extent to which these two policies can provide a solution to the unequal ramifications of the oil industry. Using the principles of eminent domain and decentralisation as analytical guidelines, the chapter shows 'who gets what' in the oil city; and demonstrates why different levels of compensation and betterment ought to be paid and received. Finally, it illustrates why and how these outcomes have not been attained.

Chapter 8 looks at the case for institutionalising more comprehensive property taxation to bring about sustainable urban development. It interrogates the idea that, because taxes can generate more income for infrastructural development and tend to curtail the syndrome of 'unearned income' accompanying greater urbanisation and public investment in urban society, merely introducing a fiscal regime is sufficient to change the distribution of gains. Based on empirical evidence collected from Sekondi-Takoradi, this chapter shows that the nature of regulation, especially the exceptions, broader economic systems of how property ought to be held, and the social and legal institutions that serve as a framework for regulations are key pre-requisites for success. Thus, the argument of advocates of property taxation ought to be revised: the need for land taxation is obvious and useful, but contingent on broader socio-economic, political, and institutional factors which are currently lacking.

Chapter 9 considers alternative uses of oil revenues. It focuses on a wide range of options and puts particular emphasis transport, mainly because of the primacy of transport development in state plans on how to utilise oil rents. It analyses statistical compendiums and oral interviews which show that the opportunity cost for using oil revenues for road construction is neglecting cleaner, safer, and more equitable rail transport. The chapter explores the historical and contemporary antecedents of bypassing public investment in the rail sector. It argues that, while investing in rail transport is a more efficient use of oil revenues, it will take sustained and fundamental change in activism and research to correct the status quo, especially where there is historical state apprehension of the railway workers' militant unionism and state appreciation of key lessons of neoliberalism dispensed by the World Bank. This chapter significantly advances the debate on oil revenues in Africa. Whereas previous studies have focused on transparency and accountability, the emphasis here is on the economic, social, and environmental efficiency of the projects on which oil revenues are spent. A strong case for policy reform is presented.

Chapter 10 concludes by knitting together key findings, reconnecting with the theory of oil cities, and making suggestions for alternative futures for theory and policy on oil cities. It recognises the impediments to change, and suggests how these may be addressed. Oil is a major force driving change, for better or worse, so understanding these political economic issues is crucial for developing intelligent and effective policies for the future.

Notes

1 For the purpose of this book, an 'oil city' is one in which oil is produced directly or whose economy is substantially impacted by oil production elsewhere, either because it provides supporting industry, services, and accommodation to the city where oil is produced directly, that the city has become home to people who used to reside in an area but have emigrated from the area of oil production where oil was produced, or that there is a rational basis to think that oil production will impact its economy substantially, because it has the infrastructure to support the oil industry. An oil city may also be one whose economy is substantially articulated with a city in which oil is produced *inter alia* such that it is a key location for spending oil rents. Thus, it is possible to talk of 'categories of oil cities'.

2 Most studies on 'oil cities' concentrate on the Middle Eastern region (Alissa, 2013; Al-Nakib, 2013; Bet-Shlimon, 2013; Damluji, 2013; Fuccaro, 2013) and rarely examine the situation in Africa (with only a few respectable exceptions such as Ali *et al.*, 2008 and Lacher, 2011).

2 Oil in orthodox theory
Repudiation and riposte

Introduction

How can we make sense of the effects of oil exploration and production on key urban issues in oil cities such as land use, employment, transport, population growth, inequality, poverty, housing, fishing, farming, and economic development? This is an important set of concerns because oil exploration and extraction are increasingly taking place in many cities around the world. Indeed, engaging with such matters has underpinned a surge in research interest in the relationship between mining, urbanisation, and poverty, culminating in the establishment of research networks such as UPIMA – Urbanization and Poverty in Mining Africa. In turn, studies are beginning to classify types of mining (e.g. Jønsson and Fold, 2011; Bryceson and Jønsson, 2012), to examine the history of mining towns and those adjacent to them (e.g. Bryceson, 2011), and to untangle issues of culture – and even of magic – for those who want to benefit from mining (e.g. Bryceson *et al.*, 2010). The effect of mining on the growth of urban settlements, urban hierarchy, and national development is also receiving sustained attention in the urban studies literature (Bryceson and MacKinnon, 2012).

Yet, the political economy of exploring and producing oil differs substantially from the mining of precious minerals. Oil dynamics more easily trigger global and local crises in food pricing, fuel pricing, and even water provision, so oil is more international as a commodity. Concurrently, oil dynamics have also been a major contributor to ecological hazards around the world (El-Gamal and Jaffe, 2010). Further confirmation of the crucial importance and uniqueness of oil are reflected in the widely held view about the biogenic origins of oil, the spread of oil over the years around the world, the geopolitics of oil, and the huge influence of the presence of oil production on the establishment of industries and other economic activities (Odell, 1997). According to two informed commentators on oil, 'No other resource dominates the world economy more than oil. Wars have been fought for control of it. Lives have been lost over it. Fortunes made from it' (Franks and Nunally, 2011, p. 3). Oil and politics are more intimately connected than the relationship between other minerals and politics and, in terms of uses, oil has far more diverse uses than gold (Falola and Genova, 2005). Indeed, the price of oil tends to drive gold prices on the international market (Le and Chang,

2012). So, oil is different from gold and other precious minerals making its study, particularly as it pertains to cities, interesting and imperative. However, to date, no theoretical framework lends itself easily to answering the chain of questions posed at the beginning of this chapter. Neither the contributions by eminent urbanists to *Theories of Urban Politics* (Judge *et al.*, 1995) nor Obeng-Odoom's (2012a) recent synthesis of theories of 'urban governance' can serve as a framework to analyse the complexities and peculiarities of oil cities.

The earliest view about oil rents and economic development, emphasising the beneficial influence of oil on the economy, was highly influential (Karl, 2004). It was related to the idea that economic surplus – defined as the difference between net national income and the essential consumption requirements of a population (Baran, 1957, pp. 22–23) – was the main driver of economic development. Following from that, the physiocrats and classical economists such as David Ricardo theorised widely about land rent and its importance to economic development (Hubacek and Berg, 2006). The neoclassical economic school provided a critique using the idea of 'resource curse' to which I return shortly.

However, none of these views is very helpful. They do not account for spatial and urban issues, the focus of this book. The existing views tend to look mainly at macroeconomic themes (Corden and Neary, 1982; Auty, 1993; Sachs and Warner, 1995; Al-mulali, 2010; Kolstad and Wiig, 2012). The sociological concept of 'anomie' is similarly inadequate because, although it looks at the 'urban', it is concerned primarily with social dysfunction, does not envisage a situation where 'oil cities' are contiguous with the oil industry (that is, oil cities developing around refineries, gas processing plants, and oil services) such that oil production does not create a 'sudden change', and has little to say about the political economic impact of oil on urban economies and environments. The political scientist's tool, the idea of a 'rentier state' (Beblawi, 1987), has similar defects, in terms of scale, unit of analysis, and predominant focus on the political sphere – although, as I will show later, the radical meaning of this concept is substantially different and key to the thesis advanced in this book.

In between these three lenses, a number of insights may be woven together to form a conceptual framework for understanding cities and oil. However, most attempts to study oil cities have focused on 'cities after oil' and looked principally at energy concerns (see Newman, 1991; Atkinson, 2007a, 2007b, 2008, 2009, 2012). The central argument of such energy-centric analyses can be summarised as follows: (1) the research on sustainable development is so grand that it forgets important, practical issues such as the future of cities after 'peak oil'; (2) the current use of energy – via the institutionalised reliance of modern economies on cars, in particular, as the primary means of everyday transport – is unsustainable; (3) the consequences of oil depletion on cities *after* oil are likely to be dire and perhaps more serious than commentators on climate change would have us believe; and (4) explicit attention to 'city futures' post energy and oil depletion are neither on the policy agenda of the world development agencies nor on the to do list of those agencies concerned primarily with cities (see Atkinson, 2007a, 2007b, 2008, 2009, 2012). Although clearly important, these insights do not provide sufficient grounding to theorise how oil reconfigures urban economic, social, and environmental development.

The present chapter attempts to close this gap. It makes the case for developing a theoretical framework of natural resources (land) and economic development that draws on Henry George's emphasis on problems resulting from the public creation, but private capture of increased site values arising from economic development; David Harvey's analysis of capital accumulation and its spatial expression; Hossein Mahdavy's analysis of the state based on the radical interpretation of the concept of 'rentier state'; and insights from Chibuzo Nwoke's analysis of varieties of economic rent. It argues that by accounting for how land accumulates rent and value, demonstrating how that rent is distributed, and examining how the state mediates the process, the work of George, Harvey, Mahdavy, and Nwoke on natural resources (land) provide important conceptual insights for understanding expectations and the ramifications of oil for urban economic development.

The rest of the chapter is divided into four sections. The next section critically reviews some of the influential theories of natural resources and economic development. The section following that presents a critical framework of natural resources and economic development, while the subsequent section shows how that theory can guide empirical research. Finally, concluding remarks highlight key aspects of the work of George, Harvey, Mahdavy, and Nwoke and shows the conceptual tokenism in mainstream theories of natural resources and economic development.

Mainstream accounts of oil and economic development

Neoclassical economists have several models to explain the relationship between resource boom and economic development. One popular explanation is derived from the 'open access exploitation thesis', which predicts that without private and formal property rights, a resources boom will inevitably lead to widespread socio-economic problems in a resource-rich economy (Barbier, 2005, pp. 122–140; de Soto, 2011). This is held to be so because public institutions and the state are assumed to lack the expertise to manage oil resources or, alternatively, that they are so entrenched in a 'culture of corruption' that they are not in a position to effectively manage oil resources. This view, widely regarded in neoclassical economics as *the* property rights approach to natural resources, was popularised by Armen Alchian and Harold Demsetz (see, for example, Demsetz, 1967; Alchian and Demsetz, 1973; Demsetz, 2002), well regarded at the time for breaking away from the established body of work on the theory of production and exchange that draws largely on a negatively sloping demand curve with only punctuated concerns about abnormal demand curves (Pejovich, 1972). So, in that sense, this property rights approach to natural resources was radical in the orthodoxy.

Another is the factor endowment framework which builds on the Heckscher–Ohlin model of international trade to argue that, through trade, sectors connected with booming natural resources sectors will gain whereas all other sectors will lose. That is, natural resource abundance in a country open to trade will worsen income distribution between those who work in the booming sector and others who work outside it (Barbier, 2005, pp. 122–140). A variant of this theory (see, for example, Wenar, 2008) states that a resources boom will invariably generate inequality without reforming the rules of trade to give greater power to resource rich communities.

A third explanatory framework, which is also the most pervasive within mainstream economic accounts, is the 'resource curse' thesis. This conception posits a negative relationship between resource abundance and economic development *inter alia* based on the assumption that natural resource abundance makes institutions corrupt, drives away initiatives of diversification, and makes the export of the endowed country uncompetitive. This paradox of being cursed by nature's gifts such as oil, diamonds, and gold became a major focus of attention in neoclassical economics in the late 1970s, but had had an earlier presence in economic discussions in the eighteenth and nineteenth centuries (Boianovsky, 2011). Then, the debate was mainly about whether it was ineffective institutions or 'effort supply' (or sloth) which caused the inverse relationship. The latter was particularly popular. Indeed, J.S. Mill is said to have observed that 'It is difficulties, not facilities that nourish bodily and mental energy' (cited in Boianovsky, 2011, p. 23). In turn, he did not look favourably on resource abundance. Following that, the 'golden age of resource-based development', from 1870–1913, ushered in the 'Staples thesis' which claimed that there was a definite positive relationship between the export of natural resources and economic development (Barbier, 2005, pp. 81–83).

Contemporary neoclassical economists now take the opposite view. A greater abundance of nature's gifts is assumed to correspond with less economic development. Two explanations for this paradox regularly put forward are issues relating to volatility and 'Dutch Disease'. The former relates to volatility of natural resource prices, which is assumed to present difficulties to endowed nations making coherent plans with resource rents (Corden and Neary, 1982). Similarly, the notion of 'Dutch Disease' refers to the tendency of the resources sector to crowd-out non-resource sectors. This is assumed to lead to an appreciation in the value of the national currency of the resource rich countries and hence makes its exports uncompetitive and imports cheap relative to locally produced goods, leading to local firms losing business and closing down.

Related to the resource curse thesis is the 'rentier state' doctrine typically employed by political scientists or economists engaging in political analysis. According to this conception, oil rents make states dependent on oil and hence less reliant on other sources of public finance such as taxation. In turn, citizens who pay no taxes become disinterested in holding the government to account (Beblawi, 1987). With channels of accountability broken, the state adopts a system of rewards based less on competence and more on nepotism in managing the economy. Such dynamics can create conflict as people become disgruntled; although, of course, conflict itself can fuel the curse too (Corden and Neary,

1982; Auty, 1993; Sachs and Warner, 1995; Collier, 2008, 2009; Al-mulali, 2010; Kolstad and Wiig, 2012).

Applied to Africa, this interpretation shades into conservative views about neopatrimonialism. The neopatrimonialism framework (for a detailed discussion, see, for example, Matter, 2010) is a method of analysis that regards, a priori, politics in Africa as enmeshed in noncorporatist personality cults where people vote for leaders because of expected personal gains, not because of ideology or broader policies. Leaders, in turn, assume office not because they intend to improve social conditions, but to attain personal wealth and wellbeing for their network of clients and associates (Bratton and van de Walle, 1994, p. 459). However, as Aaron deGrassi's paper in *African Studies Review* (deGrassi, 2008) shows, this approach has severe limitations, ranging from Eurocentrism, ahistoricism, simple binaries of 'clients' and 'patrons', and politically prejudiced stance, leading the analyst to assume that neopatrimonialism actually exists *everywhere* in Africa – a continent of 54 countries with varying histories and diverse experiences. The concept takes as a given that the 'African state', especially, is corrupt, and in need of 'help', without considering that some of these pathologies are themselves creatures of the world system (Amin, 1972, 2002). Assuming that there is a homogeneous African state, 'potted like a plant', to borrow from Francis Nyamnjoh's (2012) critical reflections on the approach, neopatrimonial analyses refuse to consider evolution, contested traditions, creation of tradition, and multi nuclei patron– client networks (see also, Pollard *et al.*, 2011).

While New Spain (Mexico) and later Norway were some of the earliest countries held to have encountered the curse, current explanations for research and policy focus have concentrated on Africa (Boianovsky, 2011), mainly because of the recent discovery of an estimated 80 billion barrels of oil in Africa (Servant, 2003; African Development Bank and African Union, 2009). Countries such as Congo, Guinea, Angola, Chad, and Nigeria are particularly (in)famous for their oil resources and curses, especially corruption (see, for example, Collier, 2009, pp. 39–44; Colgan, 2011). In turn, Paul Collier (2008, 2009), a neoclassical economist, calls the problem a 'corruption trap'. According to this view, keeping oil agreements secret from the scrutiny of broader masses, opaque expenditure of oil rents, and general corruption among governments are the main drivers of resource curse. Such poor governance practices lead to conflict, low economic growth, and growing deprivation. Dysfunctional governance breeds more dysfunctional governance which, in turn, is made endemic by low economic growth.

The posited solution to the resource curse, neoclassical economists argue, lies in providing aid – financial and human resource expertise – to African countries. Yet, Collier (2006) argues that, because of deep traps, especially in African countries, it is a 'big push', rather than just aid, that is needed. For Sachs, this 'big push' should begin with setting in place policies that ensure transparency in negotiating and executing oil agreements as well as in managing the oil rents. Next, it is necessary to implement 'process conditionalities' – that is, making it possible for the people to hold governments to account (Collier, 2006, p. 209). In turn, there is the need for financial institutions to act in such a way as to expose the financial transactions of corrupt officials (Collier, 2006, p. 207). Finally, an army external to African countries should also be put in place to police corrupt leaders, ensure that conflicts do not arise and countries remain stable after conflicts, especially when there are armed insurrections against democratically elected governments (Collier, 2006, pp. 203–205; Collier, 2009, pp. 202–212, 218–227). Together, it is posited that these policies would unleash high doses of disincentive to governments that may want to entertain corruption.

To provide a framework to avoid the resource curse, five key 'good governance' practices are typically recommended by neoclassical economists (Collier, 2006, 2008, 2009). These are Publish What You Pay (a policy which insists that transnational corporations should disclose how much they pay to governments in Africa), Extractive Industries Transparency Initiative (a framework emphasising the need to be transparent in the collection and management of oil royalties and taxes), and Equator Bank Principles (to which certain banks and institutions have subscribed and thus undertaking to give loans only to transnational organisations (TNCs) that are socially and environmentally sensitive). A fourth principle is the Wolfsberg Group principle which commits banks to anti money laundering practices. The final principle is the African Peer Review Mechanism (a process by which African governments submit themselves for peer review by other African countries on some benchmarks for governance).

Although said to be well intentioned, the five good governance principles assume that markets are efficient and 'irrational behaviour' in the market can be corrected by ever increasing regulation. Thus, the solutions seek to extend the arms of the state to ensure a better functioning of markets. They tend to naturalise the 'oil curse', assuming that oil per se inherently curses institutions.

A major drawback of the neoclassical economic framing of the oil–economic development nexus is that it is developed largely in abstraction from spatial dynamics. In particular, urban level analysis in resource curse studies is undeveloped. Recent attempts (e.g. Ackah-Baidoo, 2012; García-Rodríguez *et al.*, 2013; Andrews, 2013) have focused on the notion that oil production in Africa creates an 'enclave economy' in which the effects of oil, especially offshore oil production, are confined to that industry alone and never to the wider urban and regional area. To correct this problem, there are normative arguments about how and why oil companies should use market principles of corporate social responsibility to spread the effects of oil and help the communities while maintaining their own profits. But is the oil–urban economic development nexus really tenuous? Many empirical studies show that there is a strong connection between oil exploration, production, development, and urbanisation. As far back as 1983, Khailil Keizeiri was arguing that the rapid urbanisation in Libya was the result of the discovery of oil. Indeed, around the same time, Grill (1984) found the oil–urbanisation link in the oil-rich states of the Arabian Peninsula, and recent historical studies (e.g. Alissa, 2013; Fuccaro, 2013) show that the link was strong in the Middle East. Gilbert and Pasty (1985) established this link in Venezuela. More recent authors like V.T. Jike (2004) have shown that oil is a contributory factor to urbanisation in Nigeria.

The reasons for the oil–urbanisation nexus varies. Some (see Jike, 2004; UN-HABITAT, 2008b) suggest that oil spillage or spillage of chemicals used for mining may cause farmlands to lose their fertility and hence people making a living from urban agriculture may migrate from 'oil cities' to other cities in search of alternative forms of livelihoods as in the cases of Nigeria and Zambia. In this way, oil causes urbanisation *outside* where oil is explored. But an application of the Harris–Todaro Model (Harris and Todaro, 1970), which shows that rural–urban migration occurs because of expected opportunities in cities, would suggest that oil may also lead to urbanisation within the oil city. Research by Markusen (1978), Okonta and Douglas (2003), and Tiepolo (1996) suggests that urbanisation occurred in cities in USA, Nigeria, and the Congo respectively after the discovery of oil. In another case, the UN-HABITAT has recently found that Cabimas, a city in Venezuela, grew at an additional rate of 3 per cent because of oil production (UN-HABITAT, 2008b, p. 33). The people who migrate to cities following the discovery and exploration of oil could be from other cities, rural areas, or other countries (Markusen, 1978; Okonta and Douglas, 2003). Even in the Ghanaian town of Obuasi, where gold (not oil) is mined, 'ever since the first [gold mining] agreement was signed, workers from all parts of Ghana and indeed from many parts of West Africa have come to Obuasi to make a living' (Ampene, 1965, p. 42). Indeed, using evidence from gold mining towns in Ghana, Bloch and Owusu (2011, 2012) set out specifically to challenge the idea that resources in Africa generate an enclave economy. Gough and Yankson (2012) also provide evidence of gold-induced urban growth, what Bryceson and MacKinnon (2012, p. 514) call 'mineralised urbanisation'. In all these studies on oil, oil-induced urbanisation occurs because of the expectation of jobs in oil cities. Expectations, therefore, should be a key research focus of oil-induced urbanisation. Oil leads to urbanisation both within and without the oil city.

The two types of urbanisation, within and without the oil cities, could happen at the same or at different times. In Saudi Arabia, they happened at the same time. According to Faisal Al-Mubarak (1999, p. 31), 'while oil industrialization has influenced the location and growth of the kingdom's Eastern Province settlements, it has also boosted intensive urban growth outside the oil producing region'. Oil production also has several growth, distribution, and environmental effects in cities (see, for example, Tiesdell and Allmedinger, 2003; Atash, 2007; Fuccaro, 2013). It may generate jobs and improve facilities and lead to urban economic growth, but not all people will benefit equally. The environment may suffer, in terms of declining crop production (UN-HABITAT, 2008b, p. 130) for example.

There is a body of sociological work that tries to consider the effect of resource booms on the local and urban levels. It is dubbed the 'social disruption thesis'. It continues to be used in the twenty-first century (see a review by Lawrie *et al.*, 2011), but it received the most attention in the 1970s and 1980s (Wilkinson *et al.*, 1982), mainly because a great number of boomtowns, especially in North America, were experiencing severe social problems. Also, there was a restriction on oil importation from the Arab world, so there was interest in learning how 'home-grown' oil producers in the West would fare and, finally, because of curiosity about why, in the Third World, abundance of resources had not led to massive economic development (Lawrie *et al.*, 2011). While the studies were mostly empiricist (Albrecht, 1982), some draw on the conceptual work of sociologists such as Emile Durkheim, particularly his work of anomie which predicted that resource booms lead to a breakdown of social systems (Wilkinson *et al.*, 1982). Unlike the resource curse thesis, the mechanisms by which this breakdown occurred were held to be mainly sociological, ranging from the inability of local facilities to accommodate immigrants who follow the resource boom, an inability of local populations to mingle well with newcomers, and the problems of mixing different cultures (Wilkinson *et al.*, 1982; Albrecht, 1982). However, like the resource curse thesis, a determinist relationship is struck between boom and doom, although the degree varies between settlements of different sizes (Wilkinson *et al.*, 1982; Albrecht, 1982). Distribution, history, and institutions count for little or nothing, and the role of transnational oil conglomerates seems to be merely 'agents of development' (Lawrie *et al.*, 2011; Ayelazuno, 2013).

Recent revisionist attempts have been made by orthodox analysts to re-emphasise 'the property rights paradigm', to use the title of two of its most respected analysts, Armen Alchian and Harold Demsetz (1973). The effort is most eloquently highlighted in the book, *Oil is not a Curse: Ownership Structure and Institutions in Soviet Successor States* (Luong and Weinthal, 2010). By introducing questions of ownership structures and how they evolve over time, the authors suppose that they have made a giant departure from the neoclassical economics framework.

However, concerns about institutions, the structure of property rights, and how they emerge and evolve have always been a feature of the orthodox property rights tradition. Indeed, Armen Alchian and Harold Demsetz (1973, p. 17) observed in their germinal paper on the property rights approach that the three defining questions of the approach are:

(1) What is the structure of property rights in a society at some point of time? (2) What consequences for social interaction flow from a particular structure of property rights? and, (3) How has this property right structure come into being?

In answering these questions, restrictive neoclassical economics assumptions were retained (see, for example, de Soto, 2000, 2011). Among these, the tendency towards equilibrium in a market society, the rational individual, the profit-maximising individual, and the presence of sufficient information in all societies at all times. Markets were extolled as *the* best medium of social organisation, while normative arguments were made in favour of pricing, privatising, and marketising society, environment, and the relationship between humans and natural resources, and the environment in which they co-exist (for a critical discussion of orthodox economics methodology and approaches, see, for example, Keen,

2003; Butler *et al.* 2009, pp. 105–117; Spies-Butcher et al., 2012; Stilwell, 2012b).

In turn, they come to conclusions that defend and extend private property in natural resources and predict doom and gloom for any ownership structure that resembles, even faintly, communal or customary property rights, as Luong and Weinthal (2010) did in their property rights approach to analysing oil and economic development. To be sure, the idea that private property rights are benign has a much longer history, as we see in Richard Schlatter's *Private Property: The History of An Idea* (1951), but some economics professors defended and canonised it into a formal approach to analysis in economic science more recently. A detailed history of how this approach developed in the discipline, starting from the work of Ronald Coase, has been offered by Harold Demsetz (2002), one of the leading figures in the property rights school in economic science, if it can so be called, so I will not repeat it here.

Rather, I shall concentrate on the alternative, heterodox property rights approaches on which I draw for analysis in the subsequent chapters. These heterodox, political economy, or heterodox property rights approaches to studying resource booms and economic development retain the three questions asked in the orthodox tradition, but expand the concerns to institutional differences across space and time and broaden the definition of 'efficiency' to include distributive efficiency (Pullen, 2013). More fundamentally, they (e.g. George, [1879] 2006; Marx, [1976] 1990; Polanyi, [1944] 2001; Harvey, 2011) start from a different set of assumptions. For instance, (1) markets are inherently ineffective in managing natural resources, (2) communal and common property are not inherently open to abuse and misuse, (3) that history and institutions matter in the running and management of natural resources, (4) that the use of natural resources in capitalist societies generate different and differentiating experiences for various social groups temporally and spatially, and (5) individuals can collaborate and co-operate.

This approach sits in the broader critical political economy framework that I used in my book, *Governance for Pro-Poor Urban Development* (Obeng-Odoom, 2013b). It emphasises political economic processes, so it requires that the study of urban policies be situated in the context of broader socio-economic and environmental analysis. Not only does it emphasise evolution in historical time, distributive equity, and analysis of how institutions of varying degrees of power interact with the economic system to produce different social, political, economic, and environmental outcomes, but also it regards local history as part of global history. It does not insist all factors must necessarily be precisely estimated to be deemed 'proved' or established, particularly when the oil industry is at a nascent stage.

That approach is built on institutional political economy. It is an orientation that emphasises the economy and society as an interlocking and interdependent system. It provides alternative explanations to inform practical socio-economic policies and strategies (Stilwell, 1992a, 1992b). There is no one heterodox property rights approach, of course, so I build on my own training in land economy, urban economic development, and political economy to analyse the nature and evolution of institutions and property rights, the construction of expectations, and *material lives* of residents in the new oil city of Sekondi-Takoradi. That does not mean that 'immaterial' factors such as culture are disregarded. However, unlike cultural approaches that perceive 'social or cultural discourses' as the sole or main drivers of the economic, the approach used in this book regards the material and immaterial, cultural and economic, political and non-political processes as co-constitutive of the urban experience (Ribera-Fumaz, 2009, pp. 454–455).

So, this approach is an exercise in problematising the world of rationality, equilibrium, allocative efficiency, poor engagement with world, local history and politics, and a rejection of the zero sum gaze that pervade neoclassical economic analysis of property rights and natural resources. As the dominant property rights approach is 'wholly consistent with the neoclassical framework and is, in fact, simply a logical extension of that framework' (Randall, 1975, p. 731), this book provides a fresh perspective on natural resources and economic development. As colleague land economist C.W. Loomer (1951, p. 396) observed several decades ago,

> When an investigator sets himself to look into 'imperfections', he has apparently begun to question the traditional assumptions of perfect competition, perfect mobility of the factors, and perfect economic rationality on the part of decision makers.... One area is not like another, and that insight into economic affairs demands a willingness to inquire into the specific circumstances of a given situation.

My approach to studying this 'oil city' looks at the urban as formed by various forces, the dominant of which are the dynamics of capital accumulation. From this perspective, I do not fetishise land as a narrowly conceived 'economic' category separate from the institutions in which it is 'embedded' as it does not possess qualities that may be expressed in the formal rationality of the market. Rather, land is embedded in social and cultural institutions and norms and, thus, subject to (competing) political and moral consideration, religious beliefs, and community management (Polanyi, 2001). I further develop these features by drawing on the philosophical and empirical ideas of Henry George, David Harvey, Hossein Mahdavy, and Chibuzo Nwoke, who have extensively analysed questions of natural resources, rent, and economic development.

Heterodox property rights approach: insights from George, Harvey, Mahdavy, and Nwoke

This section develops a heterodox property rights approach to the political economy of natural resources by drawing on the ideas of four heterodox political economists. Their various approaches are considered and pursued because of their demonstrated strength at dealing with the crucially important notion of rent and economic surplus which are key to understanding the construction of expectations and unravelling the ramifications accompanying the rise of the oil industry, and broader issues of economic development. The discussion provides a strong basis to develop a framework that shifts the terms of the debate away from narrow conceptions of 'development' and the problematic notion of 'resource curse'. I begin by considering the ideas of Henry George; next, David Harvey; followed by Hossein Mahdavy, and then Chibuzo Nwoke.

Georgist philosophy and political economy

While neoclassical economists typically dismiss the ideas of Henry George, indeed sometimes calling him not a 'serious scientific thinker' (Pullen, 2012, p. 119), there is formidable evidence in contemporary political economy that his ideas and method remain relevant and needed (Gaffney, 1994, 2008, 2009; Stilwell and Jordan, 2004; Stilwell, 2011, 2012a, p. 92). George laid down his theory of natural resources (land), and economic development in his stimulating book, *Progress and Poverty* (George, 1879). He developed the theory in *The Land Question* (George, 1881) and *The Science of Political Economy* (George, 1898). As with the physiocrats and classical political economists, George put land, labour, and capital as the factors of production. He defined land as all the free gifts of nature, labour as human exertion, and capital as stock of wealth. Yet, for George, while these factors of production are all important, he placed particular emphasis on land. Labour initiates production, but cannot do so without land. Capital is important and enhances production but, without it, production is possible (George, [1879] 2006). Concomitantly, he argued that the source of growth is not capital, but labour. This suggests that, contrary to claims typically made by neoclassical economists, it is labour, acting on land that is the crucial driver of production.

To George, all the factors of production worked well together and it was not their workings that caused poverty amid progress. Rather, it is the laws of *distribution*. George argued that on its own, capital was nothing. Labour created capital, so the source of all progress in society is labour. Nevertheless, capital greatly expanded and enhanced the exertions of labour, so capital is necessary and deserving of its reward. The nature of land, George argued, is rather different. Like labour and capital, it is crucial in production, but unlike labour and capital, it is not the creation of anyone. It is a free gift of nature. In turn, the appropriation of rent – which he defined as the advantage derived from the monopolisation of nature (George, [1879] 2006, p. 91) – by a landowning group is highly problematic and indeed the source of the notion of 'progress amid poverty' problematique. As noted by George (1879 [2006], p.xvi), '[t]he fact that rent always increases with material progress explains why wages and interest do not'.

George began his thesis by looking at what drives up land values. He showed that the value of land increases with (a) population growth; (b) technological advance; (c) general advance of the society; (d) greater quality of labour *inter alia* through increased interaction and education; and (e) speculation. George stressed that, although population growth tends to increase rent through lowering or extension of the margin of production and increasing productivity (following David Ricardo) (1879 [2006], p. xvi), it is not only population growth that drives up rents. Indeed, demand expands even if population is stagnant. Individual demand, George argued, expands with the progress of society – as does rent. Similarly, the exertions and interactions of labour and the use of capital combine to increase the value of land. To George, it is not only the advances that enhance labour productivity that tend to increase land values but also auxiliary productive advances such as the activities of governments (George, 1879 [2006], p. 141). Since these factors tend to expand as society expands and progresses, landlords tend to become richer as society's economic surplus increases. Labour and capital, on the other hand, suffer with the advance of the landowning class, as wages and interest are swallowed up by rent.

Thus, the only class that benefits when land is privatised or property has been created in land is the class of landowners. George used the notion of 'progress amid poverty' to explain this role of land in economic development. It was an apt description because he observed growing human capacity and advance, and broad progress in society, measured by increasing productive power, sitting side-by-side with growing poverty and a tendency for wages to decline relative to rent and interest (George, 1879 [2006], p. xv). George argued that poverty arose not from the lack of capital, the result of nature (extreme population growth), or from nurture (poor talent, low education, low skills). Rather, poverty arose from the process of distributing wealth between land, labour, and capital. When land is privatised, landlords take the biggest share of wealth and they typically do so merely by speculation.

George charged that commodifying nature is economically and socially inefficient. He held this to be particularly disturbing because it tends to encourage speculation, is a disincentive to work, and directs land to uses that do not necessarily satisfy human need. Worse, it is at the root of the problem of 'progress amid poverty'. However, quite apart from its economic malaise, the process is morally unjust. George argued that it is only those who work that must be paid, only those who contribute that must have a share in the final produce, and only those who create and hence own their property that must be rewarded. Land, George argued, is a free gift of nature and therefore cannot be appropriated as private property.

What, then, is the way forward? George's proposed remedy is to abolish private ownership of land and make land common property. In his own words, '[n]othing short of making land common property can permanently relieve poverty' (George, 1879 [2006], p. xvii). He argued that the maldistribution of economic surplus will be corrected if land is made common property. However, George differs from others in history who have called for common ownership of land in terms of the method for ending private property in land. Unlike others who favour land redistribution, nationalisation, or government regulation, George argued that it was rent that needed to be redistributed not land per se.

To do so, George advocated the institutionalisation of a land tax. He distinguished this from tax on housing per se. Housing, George observed, was often confused with land, because of the notion of 'real estate' which combined housing and land. However, according to George, it is unnecessary to tax housing because it results from the exertion of labour. That is, taxing housing can only be immoral and a disincentive to work. From this perspective, George's concern is taxing away undeserved gains, often arising from the enclosure of land in a process where every dynamic of society adds to, rather than takes from, the value of land. Thus, by land tax, George meant tax on economic rent, which is defined as the value of land 'after all the expenses of production that are resolvable into compensation for the exertion of individual labour are paid' (George, 1898, p. 150). He argued that land tax (or tax on economic rent) meets all the principles of taxation (as developed by Adam Smith, 1776): it is certain because it is predictable; it is fair because it corrects ill-gotten wealth; and it is efficient because it is relatively easy to collect, discourages speculation, and gives incentive to produce (George, [1879], 2006).

George recommended the abolition of all taxes, apart from land tax, although modern political economists influenced by Henry George do not usually go this far. He argued that taxes such as property rate, stamp duty, and income tax should all be abolished as they unduly penalise industry. In their place, George advocated a single land tax to wrestle wealth secured through possession of nature's free gifts away from the hands of landowners. Georgists contend that the income from tax can then be invested in the provision of social goods, such as public housing, hospitals, schools, roads, and general investment in society (Stilwell and Jordan, 2004). Such public investment would increase the value of land, of course, but again, the single tax will 'mop up' all these excesses for further investment intended to drive social progress.

While George did not make an explicit connection with environmental sustainability, contemporary Georgists (e.g. Feder, 2001; Stilwell, 2006, pp. 85–92; Stilwell, 2011) have argued that an extension of the land tax argument to attain ecological sustainability is possible. The institution of land tax, they argue, is likely to discourage the misuse of natural resources because people will not appropriate scarce resources which are too expensive to maintain. However, they do acknowledge that, of itself, taxation is not sufficient to make urban economic development 'green'. Indeed, it has the tendency to create a new market for 'green' equipment that makes exhaustive use of the environment, and potentially depletes it, in order to address pressures on the environment (Stilwell, 2012). Herein lies the tension in adopting a market solution to attain ecologically sustainable urbanism.

However, land tax does have its strengths too. It can discourage speculation. If landowners have to pay a high rate of taxation, those who merely leave their land bare to accumulate value resulting from public investment in road construction and other infrastructure improvements may be discouraged from doing so because of the cost involved. In turn, the tendency for urban sprawling and hence wasting land would be curtailed as will the environmental cost of commuting between suburbia and the city centre. Moreover, the release of more land from hoarding into productive use will discourage the urge to develop and hence destroy preserved areas. Further, a tax on mineral resources that are finite can discourage excessive mining and hence enable future generations to attain some advantages from the present use of land. Indeed, a tax may be placed on present mineral rents, what Stilwell (2011) calls a 'resource rental tax', while the resulting revenue would be available for investment for present and future uses. George's argument of using common property for common purposes or for the enjoyment of all will be consistent with using environmental regulations, beyond taxation, to ensure equitable and sustainable use of minerals, such as oil, and fishing and farming rights in such a way that oil spillage on farms or in the sea or enclosures for oil drilling will not deprive farmers and fishers from losing their right to use common land. Thus, taxing the portions of the commons hitherto used for fishing and farming, but now enclosed for oil drilling and production, will be within George's environmental and ecological management frame. More broadly, if the level of equality George advocated is attained, there will be greater interest in helping one another, protecting the commons, and the aggressive pursuit of riches will decline. It is this large outcome of progress that George called the 'law of human progress' (George, 1879 [2006], pp. 275–286).

David Harvey and critical urban political economy of property rights

David Harvey, drawing on Marxian political economy, provides an entirely different diagnosis of natural resources and urban economic development but arrives at similarly profound conclusions about the production and distribution of economic surplus (Harvey, 2006a, 2006b, 2008, 2011). The starting point of Marxian analysis is investigating the nature of capitalism. Marx pointed out that capitalism reproduces itself via capital accumulation. This process of building up capital takes place in two ways, via primitive accumulation ('previous' or 'original' accumulation) and expanded reproduction (the second stage of further build-up of capital). Primitive accumulation leads to capital accumulation, but its roots are not anchored in the capitalist mode of production. Capital can expropriate resources not themselves produced by capitalist methods. Expanded reproduction, on the other hand, is glued to the capitalist mode of production (Marx, [1976] 1990, p. 873). To Marx, both primitive accumulation and expanded reproduction are organically linked and there is no suggestion that either is defunct. Indeed, Marx ([1976] 1990, p. 875) wrote of 'so-called' primitive accumulation only as 'appearing primitive'.

Harvey rekindled interest in these concepts by introducing the notion of 'accumulation by dispossession', which is effectively a rebrand of primitive accumulation (Glassman, 2006). Harvey (2003) sought to stress the ongoing and contemporary nature of 'predation, fraud, and violence', and to correct the tendency in some political economic analysis that regards primitive accumulation as temporally specific to prior stages of capitalist development (Harvey, 2003, p. 144). The concept simultaneously acts as a counter to the assumption underpinning the work of contemporary neoclassical economists that primitive accumulation is now outdated, as observed by de Soto (2000, pp. 198–199):

The expropriation of small proprietors from their means of subsistence, the access to private property rights stemming from feudal title, the robbery of common lands.... These conditions are difficult to repeat today. Attitudes have changed – to no little extent because of Marx's own writings. Looting, slavery and colonialism now have no government's imprimatur.

To Harvey, the 'old' concept of primitive accumulation led some analysts to regard capitalism within an 'inside–outside' framework, with issues such as the use of political force to create economic expansion for capital, focusing almost entirely on labour exploitation. Harvey (2003, p. 146) argues that accumulation by dispossession comes about through both co-option and confrontation – in both cases; the outcome is appropriation and further commodification.

Accumulation by dispossession thus adds a new dimension and possible source of dislocation to the capitalist economy – but also potential sources of solutions of temporal or spatial fixes. Harvey also argues that a change has occurred in the global economy with (in general) a shift away from relatively intensive methods of production which dominated in the post-Second World War period to more expansive forms of accumulation by dispossession in the era of neo-liberalism (Harvey, 2003; Dunn, 2007, p. 6). This conception holds that the processes by which capital expands are legion, but include 'the commodification and privatization of land', 'the conversion of various forms of property rights into exclusive private property rights', and 'the monetization of exchange and taxation' (Harvey, 2003, p. 145).

However, as with Marx, Harvey does not dismiss capital accumulation as entirely destructive. So, although Harvey's notion of accumulation by dispossession emphasises the expropriative *tendency* of accumulation – especially during the neoliberal era – he favours a more careful analysis of the nature, causes, and effects of accumulation in practice, as capitalism is shaped and constrained by specific factors within specific geographies. Thus, on the one hand, it is possible for windfalls to create wealth and be pro-poor – in terms of job creation and GDP growth – but, on the other hand, they can be expropriative too – for example in terms of losing fishing rights. Reconciling these would be useful, but there is formidable evidence to show that capitalism as a system is characterised by inherent contradictions.

Even apart from the contradictions in the relationship between labour and capital, ecological Marxists have pointed out that there is a 'second contradiction', that of nature and capitalism. The leading theorist on capitalism's second contradiction is James O'Connor who so became by default when he wrote his stimulating paper, 'Capitalism, nature, socialism: A theoretical introduction' (O'Connor, 1988). For O'Connor, there is an inherent dynamic in capitalism to destroy the ecological conditions for capital accumulation and hence to set in motion a chain of economic crises. While capitalism tries to lower the cost of production, it inevitably destroys the environment which is the very basis of its existence. Thus, the process of capital accumulation is one in which capitalists dig their own grave through an irreconcilable conflict with nature. Thus, to

O'Connor (1994), the idea of 'sustainable development' under capitalism or 'sustainable capitalism' is a hoax, a chasing after the wind.

The state is very important in all these analyses, both by George and Harvey. George gave a limited role to the state, namely collecting land taxes and while that view remains highly influential to this day (see, for example, Dye and England, 2009; Alterman, 2011, 2012; Fainstein, 2012), George had no theory of the state (Feder, 2001). Yet, a careful view of the state is important for the purpose of developing a framework for analysing expectations and ramifications of oil.

Marxists do better than Georgists, but their attempts at theorising the state vary widely. Indeed, a theory of the state was for a long time missing in Marxist analysis. Attempts to theorise the state within Marxian circles by scholars such as Leo Panitch in the 1970s were mainly to challenge the view within social democratic circles that the capitalist state could be autonomous. Such attempts bore fruits as the new Marxist theory of the state generated considerable interest and acceptance. While there have been some dissenters, it seems that today there is some consensus, namely that the state has to be considered as a category worthy of being studied (Panitch, 2002). In turn, a Marxist analysis will usually have one or another of a 'theory of the state'.

Usually, however, they contend that the state in capitalist society is an instrument of capital and, therefore, exists only for the capitalist class. In this regard, the state performs two important roles to enhance the accumulation of capital. First it provides ingredients such as communication and transportation services to support the accumulation process. Second, it provides or supports the institutions to contain social conflicts that may derail capital accumulation. It is these roles that O'Connor referred to respectively as 'accumulation function' and 'legitimisation function' (O'Connor, 2002). Any or both of these functions may produce outcomes that are favourable to other classes (Pickvance, 1995, p. 253). However, such an outcome would not mean that the state *intended* positive outcomes for any class, apart from the capitalist class.

In a democracy, the state needs broader 'consensus' among the populace than that offered by Marxist analysis. Hence, the state needs to be viewed not only as an avenue for oppression but as a forum of struggle between the various classes. From this perspective, the orientation of the state at any time will be dependent on the balance of forces between it, civil society, and the voting public, social movements, trade unions, and international regulatory regime. As such, the state may sometimes make concessions to enhance 'legitimation' and secure the conditions for reproduction of the political, economic and social order (Pressman, 2001; Herbert-Cheshire and Lawrence, 2002). Even Marxist political economist James O'Connor seems to concur that the role of the state in capitalist society is contingent. According to him:

> Although the capitalization of nature implies the increased penetration of capital into the conditions of production ... the state places itself between capital and nature, or mediates capital and nature, with the immediate result that the conditions of capitalist production are politicized. This means that

whether or not raw materials and labor force and useful spatial and infra-structural configurations are available to capital in requisite quantities and qualities and at the right time and place depends on the political power of capital, the power of social movements which challenge particular capitalist forms of production ... state structures which mediate or screen struggles over the definition and use of production conditions, and so on.

(O'Connor, 1988, pp. 23–24)

More recently, Harvey (2006b, pp. 25–29) has narrowed down his analysis of the state. Concentrating on particular states, he has coined the term 'neoliberal state' to describe institutions that either suppress opposition to neoliberalism or support neoliberalism by lubricating the policy sphere with market enhancing tools such as tax breaks and holidays, and creating the requisite business con-ditions to allow neoliberalism to fester. That is, the state itself can have different features under capitalism. Such features may be local, national, or international. They may sometimes contradict or complement one another (Dunn, 2009, pp. 307–312). The essence of the Marxist theory, or theories of the state, then, is to view the state broadly within prevailing empirical circumstances. Applied to this study, the question becomes what does an oil state do?

Enter Hossein Mahdavy and Chibuzo Nwoke

Iranian political economist, Hossein Mahdavy in his seminal work in 1970 laid down the concepts of 'rentier states' and 'external rents'. 'Rentier States', Mahdavy observed, 'are ... those countries that receive on a regular basis sub-stantial mounts [sic] of external rent'. External rents are, in turn, defined as 'rentals paid by foreign individuals, concerns or governments to individuals, concerns or governments of a given country' (p. 428). Mahdavy observed that, while oil rents constitute economic surplus, the state fails to turn it to pro-growth ends. In other words, there is an inverse relationship between the amount of oil rents received and how much economic growth the oil state can obtain. Mah-davy's own diagnosis of the cause was quite radical. He noted:

The explanations for this unexceptional performance may be sought in at least two different – though perhaps complementary – directions. One approach would be to focus attention on the kind of socio-political organizations that often prevail in this kind of (usually foreign-dominated) ... exporting country. It could be argued that the socio-political structure of these countries, saddled with legacies of open or disguised colonialism, is not conducive to rapid growth.... The abundant financial resources cannot be properly utilized until the socio-political barriers to growth are removed independently.... A some-what different approach would be to enquire whether additional causes may not be at work, so that even after the removal of socio-political barriers, a dif-ferent set of problems may not hinder rapid growth.

(p. 434)

Here, Mahdavy points to ineffective utilisation of oil rents by rentier states focusing on too many imports and hence a failure to develop other sectors untouched by the oil industry in what he called the 'imbalance in the input-output matrix' (Mahdavy, 1970, p. 436). In contrast, he favoured employment-generating industrialisation, noting:

> Industrialisation need not of course be the only road to rapid growth. But apart from the fact that for most underdeveloped countries industrialization seems the main hope, increasing the overall productive capacity of an economy is greatly dependent on such factors as higher capital per worker, improvement in technical skills of the labour force, greater specialization and realization of potential external economies in production.
>
> (pp. 435–436)

Mahdavy also called for 'direct subsidization of industries having employment effects or widespread inter-industrial linkages' (p. 437). Overall, he called for 'a well-planned semi-socialist state if some of the short-run and long-run deficiencies of the Rentier States are to be avoided' (p. 437). Some scholars purporting to use Mahdavy's theory have strayed into an area implying there is a 'culture of traps' and 'neopatrimonialism' in oil exporting countries. This misrepresentation or new interpretation seems to have begun with the publication of the book, *The Rentier State* by Hazem Beblawi and Giacomo Luciani in 1987 (see Sandbakken, 2006). Those following this alternative interpretation have claimed that the state in an oil economy becomes so dependent on oil rent that it no longer demands of its citizens taxes. In turn, it loses its obligation to provide services to the citizenry who also lose the 'connection' or 'power' to hold the state to account because they no longer pay taxes. Other ramifications of the rentier state is that the state becomes an arena of conflict as various aspects of it try to lay claim to the resource rents (Ukaga *et al.*, 2012, pp. 154–156).

Yet, the research underpinning the book itself was the source of harsh criticisms. According to one reviewer,

> there is sadly too much repetitiveness, expressing the same observations in different words, with hardly any progressive development of argument. Nor is one left with the impression of a rigorous research process in which problems are appropriately identified and findings appraised.
>
> (Selby, 1988, pp. 555–556)

The original Mahdavy theory was not without fault either. Indeed, a major gap in Mahdavy's theory is the neglect of analysis between actual rent received by the oil state and potential rent the oil state can but fails to extract, and how the international division of labour and local forces shape or constrain the amount and types of rent.

It was in these areas that the Nigerian political economist C.N. Nwoke made his most important contribution to the nature of the oil state. Using the metaphor of 'comprador states', he extensively analysed the notion of rent and how the state in an oil society appropriates it or shades of it. Drawing on Marxist political economy, he commenced his analysis by taking the concept of rent seriously. The notion of rent in Marxist Political Economy is discussed at length in Vol. 3, chapters 37–49 of *Capital* (Marx, 1956). Marx identified four types of rent, namely differential rent (type 1) and differential rent (type 2), absolute rent, and monopoly rent (see Harvey, 2006a; Nwoke, 1984, for more detailed commentary). Type 1 differential rent refers to the payment that accrues to landlords based on the fertility of their land. This view is in accord with the Ricardian notion of rent. Type 2 differential rent, however, differs from the Ricardian theory because, unlike Ricardo, Marx argued that fertility can be affected by the application of capital which, in turn, leads to the generation of a different kind of rent from Type 1. For absolute and monopoly rent, Marx conceived of them as rent arising because of the ability to exclude others from the use of land.

Nwoke (1984, 1986) extended Marx's theory of rent to oil production, the generation of oil rent, and the role of the state in the process of rent extraction. According to him, how oil is priced and what oil rents can be captured by the state can be understood within the Marxist theory. Thus, oil of different grades normally acquires different prices. Similarly, the ability to monopolise oil rents through, for example, economic nationalism can help to trap all the oil rents. Further, the application of capital to the oil industry can help understand how only differential rent type 1 can be appropriated by landlords in the 'Third World' that are constrained by technology and so become dependent on oil companies in the Global North. While, through economic nationalism, the Third World can appropriate absolute and monopoly rent, currently that option has not been taken. Even if it were adopted, Nwoke argues, the Third World would still struggle because any attempt to take more of the rent would lead to capitalist oil firms moving to other geographical locations to avoid losing rent to the state. Another strategy for capitalist firms to avoid losing rent is through capital accumulation, either primitive accumulation or expanded reproduction, to open up new areas or use new areas for further accumulation. Through these means, capitalist oil firms escape the brake of rent extraction imposed by their home states and move to other territories, as they did in colonies, to capture land and extract economic rent. Both strategies entail covert or overt attempts to try to control the state, explaining why oil companies try to influence the state by being economically and politically friendly – a dynamic that comprador cadre with colonial mentality struggle to appreciate. The shift to other locations and other areas where capitalist firms can exert influence give the oil companies a huge advantage to capture more rent, something Nwoke regarded as 'coloniser rent'. In his own words:

> Capitalist expansion can thus be seen as having its roots in the contradiction of two structures characteristic of capitalism: capital and landed property. Because of the problem of rent, this expansion initially tends to take the form of politically pre-emptive claims on particular new areas, for those

initial rights of landed property offer a 'colonizer's rent' that would be inaccessible to all late-arriving capital. It is against this background of rent that we can understand why trade is said to follow the flag, and why the capitalist class is particularly concerned with establishing influence over the state as the regulator, allocator, and enforcer of property rights.

(Nwoke, 1984, p. 68)

Nwoke entertained the idea of economic nationalism, a kind of regional development strategy or even nationalisation, to enable the Third World mining state to capture more rents than it currently does, but its comprador characteristics and technological dependence provide structural barriers to the success of these strategies.

More recently, Bougrine (2006) has argued that the barriers are not only erected by the lack of technology and the work of comprador cadres, but the world system itself. He argues that the close affinity between oil companies and the governments of their home countries enable them to obtain sanctions, threats of war, and diplomatic isolationism against bold Third World states that try to capture more rents from oil companies. Further, regional and international bodies claiming to be supportive of the course of development of the Global South tend to clamour to support oil companies at the expense of any attempts at changing the status quo.

It is important to consider how these ideas from George, Harvey, Madhavy, and Nwoke can be used to frame the rest of the discussion in this book. What questions will need to be asked, what data need to be collected, and from where, and how should the data be interpreted? The next section tries to address these questions.

A heterodox property rights approach: synthesis of the insights from George, Harvey, Mahdavy, and Nwoke

This book is informed by the critical *property rights* insights of George, Harvey, Mahdavy, and Nwoke. From their perspective, a study of expectations and the effects of oil prospecting, exploration, and development on the urban economy must consider the nature of urban investment and population trends in land rents (also called 'site value' or 'land value'), and effects on inequality and poverty levels. It should pay attention to the role of institutions such as the state and how it mediates the processes of extraction and economic development. The study ought to recognise the co-existence and co-evolution of crisis in the form of social deprivation, moral injustices, and economic inequality amid socio-economic progress in the form of enhanced opportunities for jobs and increased local revenues. In a strict sense, it needs to posit a struggle between capital (oil companies) and labour (local, national, other Africans, and expatriates) on the one hand, and landlords on the other hand, where the ascent of the class of landlords would appear to be at the expense of workers and capitalists. *Additionally*, the original or primary struggle between the state (representing non-oil domestic capital and labour) and the (typically foreign-owned) oil companies ought to be closely studied. That is, the analyst ought to look at the distribution of the oil rent – first, between the international oil companies and the national state and, second (once the state's share of the oil rent has been maximised), between the social classes and geographic regions, which are the key points of conflict in oil producing countries. Attention must also be given to the conditions of labour given the importance all four political economists give to this category in their analyses.

Specifically, at least three sets of questions ought to be asked. First, it must ascertain the conditions that typically improve land values in an oil economy. What is happening to investment in infrastructure? What job prospects are opening up, in which sectors, and at what levels? What are the migratory flows to and from the oil city? What is the nature of speculation? What are the changing property relations?

A second set of questions will consider effect on land rents as distinct from 'rental value' (the price for occupying housing for a specified period of time). By what margin has land and rental values increased? In what parts of the city and in what type of land (industrial, commercial, or residential) has the increase been highest?

A third and final set of questions will need to estimate how the increase in the value of land is shared among different classes. Are there visible signs of inequality in terms of housing and land ownership? What is the situation like regarding expropriation of former farming and fishing grounds/rights, as these uses are quite common in cities in Africa (Obosu-Mensah, 1999)? What evidence is there about oil spillage? Is there dispossession? What are the trends in dispossession of farming and fishing rights in the city? Is there a perception of inequality among the population in the city? Is there land consolidation? Is there land fragmentation? What taxation regimes are in place? Does the state have the capacity to tax? In what ways are tax revenues used? In what other ways can tax revenues be used, that is, what is the opportunity cost of using tax revenues in a particular way?

This heterodox property rights approach to political economic analysis, being applied to spatial and urban economic processes, ought to take seriously issues of economic surplus, its generation, control, and distribution, the activities and relationships of the state and the oil companies, socially, economically, and ecologically. Figure 2.1 is a schematic representation of how oil discovery, exploration, and production can affect urban economic development, and how the state can shape, restrain, or constrain the outcomes of such processes, from various perspectives.

Figure 2.1 is, therefore, an annotated diagram of possible ramifications of oil production for urban economic development. It does not assume that oil cities in general suffer from a complex of major social, economic, environmental, and political problems. Rather, it takes local institutions seriously and considers the activities of social movements in determining the ramifications of oil for urban economic development (Boudet and Ortalano, 2010). While as a result of the

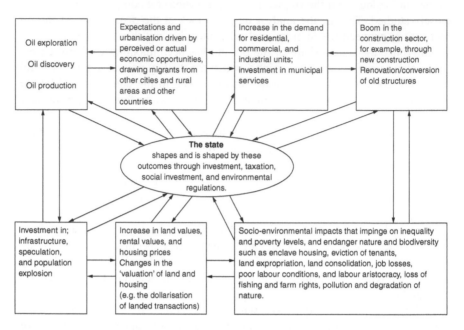

Figure 2.1 Schematic representation of a critical perspective.

discovery and production of oil, and the related investment in cities, speculation and population can increase, their dynamics depend on the institutional mix at the national and urban levels. Oil cities are likely to attract migrants, following real and expected opportunities generated by the oil industry. In turn, the demand for residential, commercial, and industrial real estate development will increase construction activities. Speculation can simultaneously drive and be driven by these dynamics, while land values will also tend to rise. Depending on whether and in what ways the state intervenes using regulations, taxation, and purposeful social investment, there could be high/low social inequality, as expressed in exploitation, expropriation, eviction, reduced job conditions, and hierarchy among workers, environmental destruction, loss in fishing and farming rights, pollution, and degradation of nature.

The contribution of Harvey will turn attention more specifically to how the processes of accumulation are dispossessing others in the oil cities and elsewhere beyond the specific drivers and forces identified by George which are more specific to the urban area. Further, Harvey's insights also point to the contradictions between different levels of spatial analysis. The data needed here generally relate to the nature of the investments of the oil companies and their profits or indirect data such as further discoveries that are being made by these companies. Further, grounded data on how different classes and social groups experience what accumulation will be needed to understand whether accumulation has been obtained at the expense of labour exploitation or land expropriation.

The role of the state will need to be investigated rigorously based on the ideas of Mahdavy and Nwoke. Data on the portion of national income that is being supported by oil rents and the nature of investments undertaken by the oil state will have to be collected to determine rentierism, efficiency, and corruption. In particular, localised ideas of the state should be expanded to consider how the state develops and its nature evolves within the global political economy. Further, data and evidence ought to be collected and led to determine whether the state is an instrument for pursuing its own interest, the public interest, or the interest of the capitalist class; and how these tendencies have come about. Further, data on external and differential rent and taxes ought to be collected among others from the contract between oil companies and the state.

Consistent with the property rights approach to political economy, all the data discussed here need to be triangulated, historicised, and internationalised. Drawing data on similar themes from different sources helps to check for consistencies and possible connections and associations, while detecting runaway estimates (Oppermann, 2000). The study must be positioned in the history of the city, using, for example, an 'institutional-analytical' method which engages historical details and records and relates them to broader social explanations embedded in political economic institutions and phenomena that are interlocking and interdependent (Tuma, 1971, pp. 3–4). As an inductive approach, it will entail critically assessing and synthesising micro and macro level qualitative and quantitative historical data, and simultaneously traversing and transcending the 'economy', the 'social', the 'environment', and the 'polity' in accounting for the 'economic' (Gay, 1930; Greif, 1998). It is this critical or heterodox *property rights* insights from George, Harvey, Mahdavy, and Nwoke, which are used for the ensuing analysis and applied to untangling the meanings and nature of expectations and ramifications of oil in Sekondi-Takoradi.

The study area, the surroundings, and how specific data were collected

Sekondi-Takoradi is Ghana's oil city. A twin city, it is only about an hour's drive from Cape Three Points off whose shores oil is drilled in Ghana. Sekondi-Takoradi Metropolitan Assembly (STMA) is the metropolis in which both cities are located. Takoradi and Sekondi are conjoined, but the metropolis is administered from Sekondi, which has been the capital of the STMA since 1946 when the two settlements were merged administratively. Figure 2.2 is a map of the STMA in which is situated Sekondi-Takoradi.

The metropolis has 27 urban settlements. Of these, Takoradi and Sekondi are the most urbanised. Indeed, they are submetros in their own right (Yeboah *et al.*, 2013). Takoradi is famous for its harbour and the proposed expansion (highlighted in Figure 2.2) is likely to make it maintain its status as more economically prosperous than its twin, Sekondi.

The current population of Sekondi-Takoradi is estimated at over 559,548 people, 51.3 per cent of whom are female (Ghana Statistical Service, 2012).

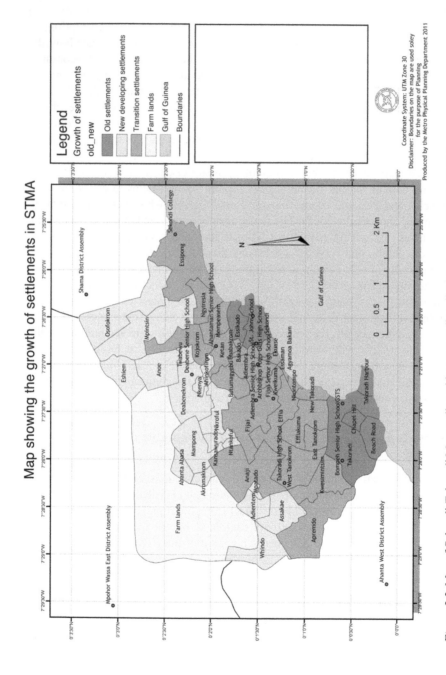

Figure 2.2 Map of Sekondi-Takoradi Metropolis and its surrounding districts (source: STMA, 2012a).

Eighteen per cent of the population are employed in agriculture. Of this share, some 5,000 people are involved in fishing and some 47,000 people farm plots of land which average five acres (Sekondi Takoradi Metropolitan Assembly [hereafter, STMA], 2011c). While about 51 per cent of the land in the entire metropolis is suitable for agriculture, the STMA (2011) has recently noted that only 34 per cent of the cultivable land in the metropolis is being farmed. The city has a huge service sector and, to quote Rémy (1997, p. 139), 'a heavy accent on industry'.

Sekondi-Takoradi is in the Western Region of Ghana. One of ten regions, it is a mineral rich area. The population of the Western Region is estimated at 2.4 million, about 10 per cent of the national population. The region is ethnically plural, but the Ahantas, Nzemas, Wassas, Sefwis, and Aowins are the predominant ethnic groups (UNDP, 2013). The Fanti language is widely spoken in the region and in neighbouring Central Region.

The urban population of the Western Region increased substantially from 36.3 per cent to 42.4 per cent between 2000 and 2010 (UNDP, 2013). Oil is a large part of this story, as we shall see shortly. Sekondi-Takoradi, the capital city, is a destination of many a migrant from other nearby settlements. Residents from Axim, Tarkwa, and Cape Three Points frequently move in and out of Sekondi-Takoradi, so the book also draws on examples from these surrounding areas. Settlements in Cape Three Points are particularly interesting because they are closest to the oil fields. These settlements are typically small in size. They include Akwida, Attenkyen, Aketekyi, Prince's Town, Egyambra, and Katakow (Yalley *et al.*, 2012) and, while they are administered by a different local authority, the settlements are strongly linked to Sekondi-Takoradi (Boohene and Peprah, 2011). Therefore, their experiences are considered as part of the analysis in this book. Besides, the close connections between Sekondi-Takoradi, its surrounding communities, the entire Western Region, its neighbouring regions, neighbouring countries (Figure 2.3), and the world at large make it imperative to consider the city as part of the Western Region but also the West African Region as a whole. The importance of considering the oil twin city in the bigger region and conceiving it as part of the surrounding settlements has been recently asserted (Yalley *et al.*, 2012). Indeed, the Economic Community of West African States (ECOWAS) has recently called on states in the West African region to adopt a regional gaze in their analysis and policy making on natural resources (ECOWAS, 2008). Therefore, I place my analysis of Sekondi-Takoradi within neighbouring settlements, as part of the Western Region, linked to the West African sub-region and African continent. That said, the main focus of the book is Sekondi-Takoradi, being the twin 'oil city'.

The primary data collected from this city during the period were of two types. The first was historical and their collection entailed visits to the Sekondi Library, the Sekondi-Takoradi Metropolitan Assembly, and the Parliament of Ghana Library in Accra (Ghana's capital city) where archival research, looking at records of past parliamentary debates on taxation, and the history of the city was conducted. These activities were complemented by in-depth interview with the

Figure 2.3 Map of Ghana, with additional annotation of some neighbouring countries and the African continent (source: Nations Online Project, 2013).

chief of 'British Sekondi' or Esikado. The second category of data related to changes in the land and real estate markets in the metropolis after the discovery of oil in 2007 and their fiscal implications. Some of these latter datasets were documented, or recorded, but not published, publicised, or analysed, so they are considered 'primary' for the purposes of this book. These primary, documented data had to be mined and pieced together from different and scattered files – some of them held in the storage of public institutions. The rest of the data were undocumented or unrecorded. To obtain all these types of data necessitated holding detailed discussions with various people in positions of knowledge on a variety of issues related to urban development, local governance, and taxation. The process of selecting respondents was non-probabilistic. It entailed a purposive sampling of people who by virtue of work experience could authoritatively comment on the research problem. These people worked for both the public and private sector organisations in the metropolis. Detailed schedules of interviews, looking at who were interviewed, how many of them were interviewed, what data were gathered and how, and the months in which the interviews were conducted are in the Appendix. In addition, I had numerous informal conversations with taxi and other commercial vehicle drivers.

The in-depth interviews that I conducted at the Ghana Tourism Authority (Western Region Division) involved three members of staff, seeking to collect data on trends in commercial and leisure real estate development. Further data on rent taxation came from interviews with three members of staff in charge of capital gains and rent tax at the Ghana Revenue Authority. To provide a sense of the investment in road infrastructure in the metropolis, discussions were held with two people at the Ghana Highway Authority (Western Region), one person at the Driver and Vehicle Licensing Authority, and three officers at the Ahantaman Rural Bank, involved in giving credit to people interested in investing in the transport industry in the metropolis. Five people at the Lands Commission, two real estate agents, and one property valuer – all of whom specialise in landed investment in the metropolis and in the land economy of the Western Region were also interviewed for data on the trends in the property market. The effects of the changes in the real estate market on landlord–tenant relationship were tracked at the Rent Control Department in Takoradi, where three people were interviewed facilitated by a discussion bearing on the activities in Takoradi with one person in the head office of the Rent Control Department in Accra because he has oversight responsibility of the Rent Control Department in Ghana. Also, following Kar (2005), who advocated the use of transect walks in doing research in the Global South, considerable time was spent doing transect walks, visiting and unobtrusively observing activities in the property market in the metropolis, but also interviewing four people in charge of various aspects of developing the largest gated housing community in the metropolis. Further, in-depth interviews with four planners at different levels of the planning hierarchy at Sekondi-Takoradi Metropolitan Assembly were conducted, supplemented by reading minutes, reports, and studying plans prepared by the planners. Finally, I carried out interviews with four railway workers, some of whom were union leaders and all of whom have been working with the sector since its vibrant days several decades ago.

To triangulate the data collected, as has been recommended for such studies by Oppermann (2000), I read the newspapers and other periodicals such as the *Oil City Magazine* which I found readily available at *Spike's Corner*. Also, I struck conversation with most of the taxi and *trotro* (commercial minibuses) drivers whose vehicles I boarded and with other passengers whenever I was travelling within and around the metropolis. Beyond this brief description, I will defer detailed data issues to chapters where they are relevant.

The third phase of the study (from March 2013 to February 2014), involved reflecting on field notes and data gathered in the light of the aims the book tries to achieve. My field notes, tape recordings, and photos – techniques whose usefulness have been well documented in the literature (Wolfinger, 2006; Tjora, 2006; Jang, 2010; Athelstan and Deller, 2013) – were very useful in this process and provided valuable data for analysis.

I went through three steps to analyse the data. They can be called data reduction, presentation, and interpretation. By reduction, I mean I winnowed the data to arrive at the most relevant. I have tried to portray the data effectively (presentation) to enable readers to verify my assessment for themselves. I have thought about the data, the process I went through to collect them, and their meaning throughout the study (interpretation). Finally, I have discussed my observations, analysis, and findings with some of the respondents and others (feedback). Taking these comments as quite representative of what others might think about the analyses, I have incorporated them in the book. This approach to analysis is inspired by its recommendation based on successful use elsewhere (Patton, 2002) and my previous experience (see, for example, Obeng-Odoom, 2010; Obeng-Odoom, 2011) with analysing field data related to urban areas in Ghana.

To further ensure validity in the analysis of the data, I have critically reflected on '*myself*' or taken self-reflexivity seriously. By self-reflexivity, I mean a process of knowledge production in which, though I am observing as a subject, I am also conscious of my position in the process. In turn, I am simultaneously a subject and object (Pagis, 2009). My emic position as a resident of Sekondi-Takoradi in the 1990s, a Ghanaian, and a writer on urban development in Ghana; and my etic perspective as a 'returnee' to the city after its new status, collectively informs my analysis. I might not fall into the trap of regarding my subjects as 'other' – a common criticism, especially of Westerners studying post-colonial countries (Spivak, 1999). Indeed, my position is strengthened by my appreciation of the city, having lived there for several years and being fluent in the local language genres in use in the twin city and, in turn, aiding me to both obtrusively and unobtrusively observe, learn, and analyse it. However, this position is not an *imprimatur* of 'natural expertise' and I risk claiming too much and presenting my own views as representative of the city, 'essentialising … ethnic identity and romanticising national origins' – a common problem for native scholars writing about their own cities and countries (Kapoor, 2004, p. 630). The problem of native researchers may also be suppressing multivocality or multi narratives and perspectives, while celebrating grand monolithic narratives (Atalay, 2008). To avoid some of these problems, I have tried to supplement and triangulate the data I collected as much as possible – a feature that can be verified by looking at the sources with which I have engaged (see the Appendix).

Conclusion

This chapter has tried to explore how we can make sense of the expectations and effects of oil exploration on urban economic development. It has argued that a heterodox property rights approach to the political economy of natural resources (land) and economic development is one effective way of identifying and assessing 'oil expectations', and analysing how the discovery, exploration, and production of oil work through multiple channels and sectors that affect urban society, economy, and environment, polity and urban development. As there is no one property rights approach, whether in orthodox economics or heterodox political economy, we have had to make a choice of which approach best suits the aims of the book.

The property rights ideas and approach of Henry George, David Harvey, Hossein Mahdavy, and Chibuzo Nwoke provide a stronger and more resilient springboard to achieve the aims of this book. So, I adopt and adapt it for the present book. Collectively, their approach shows why the tendency by neoclassical economists to draw a determinist relationship between resource boom and curse is highly problematic. Similarly, it shows why it is misleading to assume that oil boom leads to anomic conditions for every class for the state can play a role to mediate the process of accumulation at the urban level. The poor theorisation in extant studies of the importance of the state in the process of urban economic development gives the basis for discussing various ways of understanding what the state does in an oil economy. Drawing on Marxist and institutional interpretations, the conceptual foundations of the state used in this book is more nuanced than the neoclassical account that narrowly focuses on market failure and policy failure. Instead, I recognise the interplay of ideas, interests, and institutions and their diverse spatial and urban impacts on different classes in the oil twin city of Sekondi-Takoradi and its surrounding settlements. The next chapter begins to apply this heterodox property rights framework to the analysis of oil in Ghana.

3   Ghana's oil industry

Introduction

From a resource curse perspective, Ghana's oil experiences are unlikely to be different from the fate of other oil-rich African countries. However, relative to other African countries, Ghana has strong institutions which, for institutional political economists, can make a huge difference. The state in Ghana has three arms, namely the central state, the local state, and the traditional state. The central state has a strong judiciary, a strong legislature, and democratic executive which is regularly changed through generally free and fair electoral means (Gyimah-Boadi, 2009). According to the Electoral Commission of Ghana,[1] as many as 17 political parties are sometimes registered to take part in Ghana's multiparty democracy. In practice, it is the ruling National Democratic Congress (NDC) and the opposition New Patriotic Party that are the largest parties (Bob-Milliar, 2011). While Ghana is one of a few countries in Africa where ideology is explicitly stated as part of political discourse (Elischer, 2011) with the ruling government claiming to be social democratic and the largest opposition claiming to be liberal or exponent of property owning democracy, in practice, they are quite similar in terms of their policies (Obeng-Odoom, 2010, 2013a) or such ideologies are typically not understood or not the reason for people to join the political parties (Gyampo, 2012). Ghana also has a strong, vibrant, and relatively free media which ranks 41st in the world and sixth in Africa, according to Reporters Without Borders (2012).

The Ghanaian state is well known to be one of the best in sub-Saharan Africa (SSA), in terms of economic management, structural management, policies for social inclusion, and public sector management and institutions – as shown in Table 3.1.

The local state in Ghana predates independence, a point on which I shall elaborate in the next chapter, but it is the post-colonial local state that is most relevant now. That structure has been around for over two decades and operates in decentralised units totalling 216 spread all over the country. In Africa, Ghana's local government system is one of the most mature (Crawford, 2010). It has a third state, which is a traditional set up made up of chiefs – an institution which has considerable respect among the population and full recognition in the

Table 3.1 Country and Policy Institutional Assessment (CPIA) 2012

Indicator	Components	Ghana	SSA IDA* average
Economic management	Monetary and exchange rate policy, fiscal policy, and debt policy	3.5	3.4
Structural management	Trade, financial sector, and business and regulatory environment	4.0	3.2
Policies for social inclusion	Gender equality, equity of public resource use, building human resources, social protection and labour, policies and institutions for environment sustainability	4.0	3.2
Public sector management and institutions	Property rights and rule-based governance, quality of budgetary and financial management, efficiency of revenue mobilisation, quality of public administration, and transparency, accountability, and corruption in the public sector	3.7	2.9
Overall CPIA score		3.8	3.2

Source: World Bank, 2013.

Note
* International Development Association, a credit-giving sector of the World Bank.

Constitution of Ghana. While the traditional state has no direct power over the central and local states and it is, in fact, barred from taking part in party politics (see Article 276 (1)), the chieftaincy institution has major powers in terms of land management (see Articles 267 (1 and 6)). Over 70 per cent of land in Ghana is held by the traditional authority constituted by chiefs, traditional priests, and families (Abdulai, 2010, p. 138).

Generally, these states co-exist peacefully and each tends to support the process of economic development. In those economic terms too, Ghana has an enviable position in Africa. Its economy is booming. On a real basis, it grew by 7.7 per cent in 2010, and, grew further at 14.4 per cent in 2011. In 2012, there was a drop to 7.9. While actual figures for 2013 are not yet available, provisional figures show that the real GDP growth rate (7.4 per cent) is still high in African (4.9 per cent in sub-Saharan Africa) and global (3.2 per cent) terms. Year-on-year inflation fell from 20.7 in June 2009 to 8.6 per cent in 2011 and rose only marginally to 8.8 per cent in 2012 and then to 13.1 per cent in October 2013 (Ministry of Finance and Economic Planning, 2011, 2012, 2013). The unemployment rate dropped from 8.2 per cent in 1999/2000 to 3.6 per cent in 2006/2008 (Ghana Statistical Service, 2000, 2008) and, during the years for which statistical information is available (1991/1992, 1998/1999, and 2005/2006), the incidence of poverty dramatically reduced from 36.5 per cent to 26.8 per cent, and then to 18.2 per cent respectively (Ghana Statistical Service, 2007, p. 9).

There is also considerable civil society oversight in the oil industry in Ghana. Prominent examples are the Western Region Development Forum, Integrated Social Development Centre, and Institute of Economic Affairs (see Heller, 2013; UNDP, 2013, for a discussion of their activities). While such organisations were sidelined initially, they were subsequently involved in most major activities (Gyampo, 2011), including during nationwide consultations to formulate the Petroleum Revenue Management Act (Amoako-Tuffour and Ghanney, 2013). Grave problems detected at the bill stage during parliamentary debates vindicate Gyampo's (2011) analysis that the civil society groups did not have very detailed understanding of the oil and gas sector. However, recent evidence shows that the problem of capacity has greatly improved. In the words of one researcher:

> Ghana benefits from a sophisticated network of civil society groups that have developed expertise and effective techniques over years of advocating for more transparency in the mineral sector and are putting it to use to promote responsible management of Ghana's new oil bounty. Ghanaian civil society organizations have conducted sophisticated analyses that have contributed to the promulgation of a forward-thinking system for the management of petroleum revenues, and are now participating in the institutions set up to implement that system.
>
> (Heller, 2013, pp. 90–91)

Currently, over 115 civil society groups are involved in scrutinising oil deals through membership of the Civil Society Platform on Oil and Gas, formed in March 2010. Furthermore, there are over 150 private FM radio stations, 20 TV stations, and some 114 internet service providers, all of which have shown interest in the management of oil revenues for the benefit of the majority of the people in Ghana (Gyimah-Boadi and Prempeh, 2012; Kopiński *et al.*, 2013). Most of these outlets tend to focus on neoliberal policies and reformist strategies extolled in the new public management literature and have succeeded in further legitimising the so-called importance and benefits of neoliberalism (Ohemeng, 2005). Together with other national and global influences, therefore, they have succeeded in making the idea of 'private sector-led' development the main focus of public policy in Ghana (Obeng-Odoom, 2013b). Nonetheless, existing evaluations (e.g. Kopiński *et al.*, 2013) suggest that they do provide a critical oversight role to hold the state in check.

For all these political economic reasons, Ghana has been regarded in the literature as one of the big four in Africa (Naudé, 2011). While the other three countries in Africa, namely Botswana, Mauritius, and South Africa, may enjoy some of these features of Ghana's political economy, what makes Ghana a particularly interesting case for study is that it has only recently discovered oil and such praise continues to pour in six years after oil discovery, production, and development. Indeed, only recently, Ghana was praised by David Cameron for its excellent economic growth record and enviable performance as 'an island of peace and stability' (*Voices of Ghana*, 2013).

This chapter provides a diachronic and synchronic view of the oil and gas industry and what role the state has played in it. It problematises existing binary accounts of Ghana's oil being a blessing or a curse, showing that there are variegated experiences that do not neatly fit established categories. The notion of rentier state is supported, not because of poor economic management. Indeed, while between 2007 and 2012, there was a decline in Ghana's CPIA from 4.0 to 3.8 (World Bank, 2013), this performance, five years after oil, is impressive by African and global standards. Compared with other resource-rich countries in SSA, Ghana's CPIA of 3.8 is much higher than other countries in the sub-region for which the average CPIA is about three (World Bank, 2013). The rate of economic growth has also been high, again by African and global standards.

However, the Ghanaian state can be called 'rentier' in the sense that oil rents have increasingly become a part of its national revenues. Even then, there has not been the open hostility of the central state to outside scrutiny, as pertains elsewhere, and massive corruption either in the oil sector or aligned sectors has been relatively minimal. The media landscape has continued to thrive and open commentary on oil production and its consequences is encouraged and vibrantly done in the media, as we shall see in Chapter 5. The state does not consistently pursue the public interest but neither does it always pursue its own interests. Aspects of both tendencies can be seen in its role in the oil era. Evidence of a propensity to systematically enforce particular class interests is inconclusive, at least at this stage.

The rest of the chapter is divided into four sections. The first section briefly sketches the origins of and global interest in the oil and gas industry. The second section investigates the prospects and challenges, while the third analyses the uncertainties surrounding the industry. The final section concludes this chapter by highlighting the key arguments, and drawing attention to the gaps there are in our knowledge of Ghana's oil industry as a segue to the next chapter.

Origins and global interest

Records from the Environmental Protection Agency (see Environmental Protection Agency, 2011c) show that Ghana's hydrocarbon industry originally emerged in 1896. Between then and now, it has gone through four phases. The first, spanning 1896 to 1969, entailed observation for onshore oil prospecting. It was significantly limited by the absence of advances in geology. In turn, the first reported and recorded discovery of oil came up to only five barrels of oil per day and this was between 1896 and 1897. Phase two, the 1970–1984 period, differed markedly from phase one, at least to the extent that it witnessed the commencement of offshore oil exploration. The first oil well in this epoch was drilled offshore Saltpond in the Central Basin. While 37 wells were drilled in all, only two of them resulted in oil finds in 1970. Fourteen years later, the PNDCL 64 (1984) was enacted to usher in the Ghana Petroleum Company (GNPC), incidentally ending the second phase of oil activities in the country. The first major achievements of the GNPC characterise phase three (1985–2000) of the oil industry in

Ghana. During that time, GNPC acquired, procured, and interpreted, for the first time, 3D seismic data of the South Tano Field. Later, it successfully drilled three wells over the Tano Field between 1991 and 1994. The fourth, and current stage of the oil industry, spans 2001 to date (Environmental Protection Agency, 2011c).

This era has witnessed dramatic scientific advances but also political and legal tensions. The oil industry has witnessed a move from shallow water drilling (0.200 m) to deep water drilling (over 200 m). Further, four deep water wells were drilled between 1999 and 2003 (Environmental Protection Agency, 2011c). Around that same time (2002), the controversial Tsikata case broke out. The details of the case were that Tsatsu Tsikata, a former boss of the Ghana National Petroleum Company (GNPC),

> In or about February, 1993 in Accra in the Greater Accra Region wilfully caused Ghana National Petroleum Corporation (GNPC) to guarantee a loan of FRF55,000,000.00 from Caisse Francaise de Development to Valley Farm, a private company which loan Valley Farm failed to repay resulting in the GNPC repaying the loan and thus causing the State to incur a loss of the said amount.
>
> (cited in the ruling of Ampiah, J.S.C., *Tsatsu Tsikata* v. *Attorney-General*, 20 March 2002)

There is extensive documentation on the issues leading up to the case, the case itself, and the fallouts from the case. Statements issued by both critics of Mr Tsikata and his own replies have recently been re-published in the well-respected newspaper, the *New Crusading Guide* (2013), edited by Ghana's famous investigative journalist, Malik Kweku Baako Jr. The case itself lasted six years (2002–2008) with many twists and turns, including the need to determine whether one of the courts in which Mr Tsikata was standing trial was constitutional. Mr Tsikata was eventually found guilty of the charge of causing financial loss to the state. The outcome of the case drew varied reactions about the independence and fairness of the judiciary in Ghana (Sah, 2008), but another way of looking at it is probably to argue that the judiciary set a strong anti-corruption example, especially when an independent jurisprudential analysis of the merits of the majority ruling in the Supreme Court (see Bimpong-Buta, 2005) would suggest that Mr Tsikata did have a case to answer.

The key point for now, however, is that the efforts of Tsatsu Tsikata at GNPC coupled with that of others contributed substantially to developing the oil industry. Indeed, in 2007, the cumulative effort of the country finally bore fruits: oil was discovered in commercial quantities in Ghana for the first time. While there is investment to develop the industry on shore to cover a landmass that spans the Volta Region, Brong Ahafo, and the Northern Regions, Ghana's oil industry is mainly located off the shores of the Western Region of Ghana. The largest oil field is called 'Jubilee', which is some 60 km from shore. The entire industry can be divided into three, based on the nature of operations: exploration; field development and production; and decommissioning and abandonment (Environmental Protection Agency [EPA], 2011c). To date, the largest oil field is the Deepwater Tano area or the Jubilee Field. According to the Petroleum Agreement for Deepwater Tano, the Jubilee Field Partners have a 30-year lease agreement, starting from 2006, to be involved in the exploration, production, and development of oil in the Deep Tano basin.

Table 3.2 shows that, while a few of the major players in the oil industry have subsidiaries in Ghana, most of the players involved in exploration are headquartered overseas but their local impacts are enormous.

Tom McCaskie (2008) was the first to raise the issue of global interest in Ghana's oil industry. In particular, he showed how the United States, using the discourse of promoting USA–Ghana security relations, is keenly interested in securing its share of Ghana's oil under the auspices of a military command for Africa (AFRICOM). Although the influence of rising global powers such as China has not been as much felt as the USA in Ghana's oil industry, by financing infrastructural projects such as energy-producing dams (e.g. the Bui Dam) and providing so-called developmental loans within the context of the history of China's engagement with Africa (which is heavily skewed to the pursuit of energy resources on the continent), China is also widely regarded as an interested party in Ghana's oil (Rupp, 2013). With Ghana–China trade relations increasing by 27 per cent in 2010 and totalling \$2 billion, China is predicted to become one of Ghana's biggest trading partners, including in oil (Aboagye, 2011).

China's engagement with Ghana has been via the use of state institutions unlike other Euro-American states that work more indirectly to support private companies from their countries (Rupp, 2013). In turn, the state marches its might, politically and economically, against private enterprises jockeying for

Table 3.2 Some players in Ghana's oil fields

Oil field	Players	Headquarters
Jubilee	Tullow Ghana Ltd	London, UK
	Kosmos Energy	Hamilton, Bermuda, UK
	Anadarko Petroleum Corporation	Texas, USA
	GNPC	Accra, Ghana
	Sabre Oil and Gas	Derrick City, Pennsylvania, USA
Tweneboah	Kosmos Energy Ghana Ltd	
	ENI Ghana	Accra, Ghana (main office, Rome, Italy)
	Tullow Ghana Ltd	Accra, Ghana
	Hess Ghana Exploration Ltd	Accra, Ghana (Main Office, New York, USA)
Odum	Oranto Petroleum Int. Ltd	Sutton, UK
	Tap Oil Ghana Ltd	Accra, Ghana (Main Office, Perth, Australia)
	Vanko Ghana Ltd	Accra, Ghana (Main Office, Houston, Texas)
	Afra Energy Ghana Ltd	Accra, Ghana (Main Office, London UK)

Source: Public Interest and Accountability Committee (PIAC), 2012.

positions in the oil industry. According to the Ambassador of China to Ghana, '[w]e would not hesitate to operate in Ghana's new oil industry if the government gives us the opportunity'. On his part, then President of Ghana said, 'China is only doing what it does best to help us … such investment is very much appreciated' (Aboagye, 2011, p. 11). Indeed, it is Sinopec, the China International Petroleum Corporation, which has, since November 2011, been heading a consortium contracted to construct Ghana's $750 million Gas Processing Plant. The financing of the project is also from a $3 billion loan offered by the Chinese Development Bank (Public Interest and Accountability Committee, 2013). While, as we shall see, the loan is secured by current and future oil revenue, there is no explicit evidence that Beijing influenced the Government of Ghana to mortgage its oil reserves, even if the mortgagee played an active role in determining the use to which the loan must be put (Mohan, 2013). There is explicit collaboration between the Ghanaian government and the Chinese state in the way in which Ghana's oil industry runs.

Ghana seems to have benefitted from the Chinese involvement. Legitimate questions have been asked about the effectiveness of the processes of environmental impact assessment that gave the 'all clear' for the work on the dam to proceed. Yet, there are also major benefits, not least the 400 MW of electricity that is badly needed by the Ghanaian society in a time that it struggles with reliable power supply (Mohan, 2013). Also, Chinese involvement in the oil-gas-energy complex has led to the generation of 3,500 jobs in the process of constructing the Bui dam (Rupp, 2013). I shall return to the issue of job creation and broader environmental–economy interactions in Chapters 4 and 5. For now, it will suffice to say that Ghana's experiences with oil signal a strong, but invisible, US–China politicking that shades into the politics of oil in Ghana.

However, it is not only China and the USA that are interested in oil in Africa: most industrialised countries are seeking to move away from the Middle East-centric focus in their pursuit of oil supplies (Mohan and Power, 2009). Since 2007, there have been at least 41 companies, from different countries, that have applied for prospecting licences to operate in Ghana. The approach adopted by the Government of Ghana in dealing with these applications is to negotiate with individual companies rather than use a competitive tendering process. The companies prospecting in the Jubilee fields and their respective shares are Tullow Oil (34.7 per cent); Anadarko Petroleum (23.49 per cent); KOSMOS Energy (23.49 per cent); Ghana National Petroleum Corporation (13.75 per cent), Sabre Oil and Gas (2.81 per cent), and the EO Group (1.75 per cent) (Akli, 2010), which sold off its shares to Tullow in May 2011. Therefore, as of the time of writing, the share of Tullow Oil Plc in Jubilee Fields is 36.45 per cent (Public Interest and Accountability Committee [PIAC], 2012). According to the agreement between the Government of Ghana and the Ghana National Petroleum Corporation and Tullow Ghana Ltd, Sabre Oil and Gas Ltd, and Kosmos Energy Ghana made on 10 March 2006, the partners must collectively pay 5 per cent of the gross production of crude oil as royalty, although the royalty for crude with the American Petroleum Institute (API) of less than 18 degrees is pegged at 4 per cent. Also, the rate of royalty for the production of natural gas is 3 per cent. The state is permitted to take cash in lieu of receiving crude as royalty, if it gives the Jubilee partners 90 days' notice (I shall return to the issue of revenues in Chapter 8).

In the lead up to the sale of the shares of the EO Group, there were attempts to marginalise the EO Group, a company owned by George Owusu and Kwame Bawuah-Edusei, because of its close connections with the former government formed by the New Patriotic Party. The Mills government (2008–2012) believed that the EO Group obtained its interest in the oil industry without following due process (*Daily Guide*, 2009). However, 'independent investigations' by *Afrikan Post* (2009) shows that the EO Group did not do anything untoward: its application was subjected to the requisite checks and approved by the Ghana National Petroleum Corporation Board and the Ministry of Energy for satisfying the requirements of the Ghana Petroleum Law. Furthermore, the appointment of Kwame Bawuah-Edusei as Ghana's Ambassador to the UN (August 2004) and the USA (September 2006) took place *after* the group's application had been approved and signed in July 2004.

The Mills Government had other concerns with the agreements entered into by the former government. On 28 June 2010, KOSMOS Energy sought the consent of the government to finalise a sale and purchase agreement with Exxon-Mobil (Oteng-Adjei, 2010). The government (represented by the Ghana National Petroleum Corporation) claimed that KOSMOS Energy leaked confidential information to ExxonMobil and suggested that action made the agreement to sell KOSMOS' interest to ExxonMobil inappropriate. It argued that, being a partner to the consortium that discovered the oil, it reserved the right to be given the first option to buy KOSMOS' stake (*Joy Business*, 2009). In the brouhaha about US interests, Chinese National Offshore Oil Corporation was keen to support the Ghana side to outbid the US company, ExxonMobil. China provided a total of $7 billion, made up of $5 billion in cash settlement and $2 billion in concessionary loans, to shore up the financial ability of GNPC to capture KOSMOS' share – an amount way in excess of the $4 billion ExxonMobil was offering. Yet, KOSMOS decided not to sell its shares. It was later to be seen trading on the New York Stock exchange under the name 'KOS' (Rupp, 2013). In spite of this abortive attempt to undo the perceived 'wrongs' of the previous regime, the Government of Ghana was relentless. It made another attempt to terminate the contract of the Norwegian Oil Giant, Aker ASA, which obtained its licence in November 2008 because, contrary to section 23 (15) the Petroleum (Exploration and Production) Law, it failed to make a local representative the signatory to the Petroleum Agreement. However, article 25 of the Petroleum Agreement makes it possible to correct that 'legal defect' by assigning its rights in the agreement to its local representative, AKER Ghana Limited, a Ghanaian registered company (*New Crusading Guide*, 2010a, 2010b).

At the time of writing, the explicitly political issues related to prospecting companies seem to have quelled. It is the resulting socio-economic issues framed around the resource curse doctrine that require further political economic analysis.

Prospects and challenges of oil production at the national level

Ghana's oil fields are mainly offshore, although LukOil has invested $100 million to support onshore oil exploration which is said to cover about 40 per cent of the landmass of Ghana, specifically in the Volta, Brong Ahafo, and Northern regions (Ghana News Agency [GNA], 2008).

According to the Public Interest and Accountability Committee (PIAC, 2012), established *inter alia* to track the management of oil revenue in Ghana, the Jubilee field has proven oil reserves of 800 million barrels of oil or potentially three billion barrels of oil reserves. These reserves make Ghana the seventh largest producer of oil in Africa (Kopiňski *et al.*, 2013). Ghana's oil is said to be 'sweet' (low in impurities and sulfur) and 'light' (easily convertible to gasoline) (McCaskie, 2008). According to the American Petroleum Institute (API) gravity, a measure of how heavy or light the weight of oil per unit volume is, Ghana's oil is estimated at 37.6 degrees (that is, API gravity of 37.6), implying that its demand among refineries around the world is likely to be high (Boye, 2010). The first 650,000 barrels of oil obtained from the Jubilee Oil Fields was sold for over $90 per barrel in the world market (Smith-Asante, 2011).

In the last three days of November, 2010, production in the Jubilee field averaged 24,395 bpd (barrels per day), but production picked up in December, averaging 37,932 bpd, and since then nearly doubled to 64,000 bpd. Overall, the total oil lifting by the end of 2011 was 24,451,452 barrels (PIAC, 2012). By mid-December, the oil production had increased from the 2011 average of 64,000 bpd to around 90,000 bpd and, by the end of December, oil production was in excess of 105,000 bpd (*Daily Graphic*, 2012). The total volume of oil produced in 2012 was 26,351,278 barrels (Ministry of Finance and Economic Planning, 2013).

Estimates of 'local content' in the supply of goods to the oil sector and job creation seem positive. For instance, 40–60 per cent of the jobs to be created directly from the Jubilee 1 phase were predicted to go to Ghanaians (World Bank, 2009, p. 22). The impact of oil on the macroeconomy of Ghana is substantial. Actual real GDP growth rate for 2011 was only 7.5 per cent without oil, but *with* oil, it jumps to 14.4 per cent (Ministry of Finance and Economic Planning, 2012). Table 3.3 provides further evidence of the growth effects of oil between 2011 and 2013.

Table 3.3 Growth effects of oil, 2011–2013

Real GDP growth	With oil	Without oil
2011	14.4	7.5
2012*	9.4	7.0
2013*	8.3	7.0

Sources: Ministry of Finance and Economic Planning, 2010, 2011b, 2013.

Note
* Projected.

In future, oil is expected to bring in about $1 billion in taxes from the Jubilee field alone (Akufo-Addo, 2010). From 2010 to 2015, oil revenue is predicted to constitute an estimated 4 to 6 per cent of GDP (Dagher *et al.*, 2010). Some of these figures are provisional, so we need to be cautious in interpreting them.

However, there are other reasons to argue that the macroeconomic effect of oil has been substantial. First, there was an 8.9 per cent increase in the total volume of oil produced between 2011 and 2012 (Ministry of Finance and Economic Planning, 2013) in a period during which world price of crude oil is generally positive. Second, the 2013 Budget of Ghana (Ministry of Finance and Economic Planning, 2012, point 48) reported a substantial total 2012 oil lifting receipts of US$541.07 million (GH¢978.27 million) of which royalties amounted to US$150.64 million (GH¢272.37 million); state carried and participating interest was US$390.43 million (GH¢705.91 million); and surface rentals and SOPCL royalties totalled US$552,418 (GH¢1,044,290). The full figures for 2013 are not yet available, but revenue from oil between January and September 2013 amounted to US$533.86 million (Ministry of Finance and Economic Planning, 2013).

It is expected that the country will produce gas from its oil at a rate of 1,000 cubic feet of gas per barrel of oil. GNPC expects 120 million standard cubic feet (mms cfd) in the short run and 240 mms cfd in phase two from the industry. Two types of gas are expected, namely gas condensate and natural gas. Thus, the three primary products from Ghana's oil industry are crude oil, natural gas, and gas condensate (Environmental Protection Agency, 2011c). Assuming a peak phase one production from the Jubilee field alone, Ghana could produce 120 million cubic feet of gas per day. Further assuming current world market price for natural gas liquids (NGL) of $2 per thousand cubic feet, the country is expected to obtain gross revenues of approximately $260 million per year in addition to the oil revenues. From these figures, a 50 per cent equity ownership in the gas infrastructure by the GNPC, Ghana's power production company, could lead to corporation tax revenues of about $120 million per year for the Government of Ghana (World Bank, 2009, pp. 2–3). GNPC records show that Ghana has a total of 159 trillion cubic feet of natural gas, making the country the second largest holder of gas in Africa (Kopiński *et al.*, 2013). When fully developed, the gas industry will save the country at least $1 million every day (*Chronicle*, 2013). Natural gas can generate many other uses and lead to the establishment of subsidiary industries, so the prospective benefit of this resource is more than its intrinsic value.

In addition to gas, there are many other potentially positive multiplier effects of oil. The World Bank has approved $38 million credit to the Government of Ghana to be used to implement an Oil and Gas Capacity Building Project. The project is intended to strengthen the technical expertise of the staff of key state institutions, such as the Ministry of Energy, GNPC, and the Environmental Protection Agency (EPA) as well as the Ghana Revenue Authority, the Extractive Industries Transparency Initiative Secretariat, the Attorney General's Department, and the Economic and Organized Crime Office. Also, the World Bank has offered $2 million in grant to empower grassroots and community participation under its Governance Partnership Facility. The declared intention is to finance the activities of civil society and community based organisations (*Daily Guide*, 2010). The government has already offered Ghana's premier university of science and technology a sum of $500,000 to build oil research capacity for studies on oil (Ministry of Finance and Economic Planning, 2012).

Not all is well with the industry. Most of the laws about oil were enacted over two decades ago when Ghana did not have oil in commercial quantities, so there are fears that most of these laws are not relevant to the current situation. For instance, Cavner (2008) has argued that the Fundamental Petroleum Policy of Ghana is too vague about how the oil industry can be regulated. Most of the existing laws, such as the Petroleum (Exploration and Production) Law of 1984 (PNDC Law 84) and the Petroleum Income Tax Law of 1987 (PNDC Law 188), were promulgated in the 1980s when Ghana was governed by a military dictatorship, so they were not subjected to broad public scrutiny, parliamentary debates, discussion, and popular approval (Gary, 2009).

These issues have several implications. First, they open up the possibility that the oil companies may act in ways which are socially suboptimal (Adu, 2009). Second, they provide a basis for avoidable litigation. Take, for example, the oil activities in Ghana's exclusive economic zone (200 miles from shore), for which there are no clear laws. Ghanaian laws are applicable to its territorial sea (12 miles from shore), a subset of the economic zone; however, sections of the zone are covered by Public International Law (e.g. the Law of the Sea) rather than local laws (Allan, 2009). As such, questions could be raised about whether Ghana has exclusive rights over portions of the oil field. Indeed, officials from Côte d'Ivoire, which shares border with Ghana, have made claims to this effect. According to them, Côte d'Ivoire should properly be regarded as a part owner of 'Ghana's oil field' in the Western Region because there are no clear legally binding boundaries that separate the territorial waters of Côte d'Ivoire and Ghana in that region. This situation brings to mind the conflict which erupted between Nigeria and Cameroon over the Bakassi Peninsula. As such, public opinion pressured the Government of Ghana to adopt every possible 'friendly' and 'diplomatic' measure to nip a potential conflict with Côte d'Ivoire in the bud (Koomson, 2010). The government made some effort in this direction.

However, resolving the problem was difficult. By law, the Government of Ghana could appoint a commission of experts to lead such negotiations with Côte d'Ivoire only if there was another legal basis on which the commission could be established. No such law was in place. Therefore, under a certificate of urgency, the Ghana Boundaries Commission Bill was sent to parliament, discussed, passed into an Act, and assented to by the president to become law. This law has now paved the way for Ghana to negotiate with Côte d'Ivoire about the respective interests of the two neighbouring countries in the oil field (Koomson, 2010). This effort appears to be working as a recent discovery in Ivorian waters did not raise any lasting tensions. In any case, there is now a firmer legal basis for further discussion and negotiation.

Still, some observers (e.g. Cavner, 2008; King, 2009; World Bank, 2009, pp. 27–30) have expressed worry about the insufficient number of institutions and civil society organisations which are available to scrutinise the activities of oil companies and government officials. They contend that, in a country with high illiteracy levels, the absence of these institutions may facilitate corruption. Ghana's decentralised system can help to reduce the likelihood of corruption which emanates from excessive centralisation. While as noted at the beginning of this chapter, Ghana has a mature and established decentralisation, Crook (2003) has found that it is administrative, rather than political or fiscal, decentralisation that is effective in Ghana.

Uncertainties

In the meanwhile, there are a number of uncertainties about the industry, especially around the notion of a 'Dutch disease'. As earlier explained by Corden and Neary (1982), this concept refers to the situation in which a boom in natural resources leads to a contraction in the non-booming part of the economy. It arises because of a fall in overall labour supply (as households choose more leisure) and a shift of labour from the non-booming to the booming sector. In turn, the output in the non-booming sector declines and the price of the factors of production in the booming traded sector drops relative to prices of factors in the non-booming sector. This aspect of the Dutch disease is called the 'resource movement effect'. There is also a 'spending effect'. Expenditure of oil resources impacts on the value of the local currency which, in turn, may affect exports and imports. Exports of the oil country may become too expensive to the point that it may cause their demand to fall, while imports into the oil country may be cheaper than locally made goods, such that there may be a rise in demand for imported goods and a fall in demand for locally produced goods. Consequently, local or indigenous industries may fold up.

The 'disease' was first discovered in the Netherlands where the discovery of natural gas in the 1960s led to negative effects on the Dutch economy: the guilder, the Dutch currency, appreciated. As a result, Dutch exports became relatively uncompetitive. Also, national attention was given to the development of the natural gas resource with adverse effects on manufacturing industries. In 1977, *The Economist* periodical coined the term 'Dutch Disease' to describe the fate of the Netherlands (Goodman and Worth, 2008, p. 204; Collier, 2009, p. 39). Since then, the expression has become part of the lexicon in resource economics.

Not many studies have been conducted about possible Dutch disease effects of the oil industry on the Ghanaian economy. Some analysts, such as the World Bank, believe that, given the existing conditions in Ghana, namely the fact that the size of Ghana's industry is small – only about one-eighth that of neighbouring Nigeria – as well as its active and continuing promotion of other commodities such as cocoa and gold, it is unlikely that it will suffer Dutch disease. According to the Ghana Country Director of the World Bank, 'It's a bit of oil, not a whole lot, so it's not enough to give you the Dutch disease and a curse'

(*Daily Graphic*, 2010a). More recently, a similar stance, but based on an appraisal of matters of governance and institutions, has been presented by Kwaku Appiah-Adu (2013). Nevertheless, as we shall see later in the book, this sanguine view is only part of the story.

The work of Dagher *et al.* (2010) remains the only systematic analysis of a possible resource movement effect. Employing a Dynamic Stochastic General Equilibrium (DSGE) model, it assumes large 'learning by doing effects' and high substitutability of labour between the non-traded and traded sectors. It further assumes a direct relationship between the traded sector and overall productivity in the economy. The study found that, although there may be some output losses in the traded sector, they are small. However, even if they are large, they are not large enough to offset the overall positive impact of the oil industry on the Ghanaian economy.

Others, such as the Centre for Policy Analysis (CEPA) in Ghana (CEPA, 2010, 2011), have alluded to the possibility of a Dutch disease scenario, particularly in the form of spending effects. The CEPA study contends that, as more revenues from oil become available, the value of the local currency will increase. In turn, imported goods will be cheaper and hence make people switch from the purchase of goods made by local producers. On the other hand, exported goods may become more expensive, leading to a drop in the demand of exports from Ghana. Overall, CEPA (2010, 2011) argues that there could be job losses, as a result of oil exploration.

These opposing views seem to leave open the question of whether there will be Dutch disease effects. Breisinger *et al.* (2010) argue that it is an empirical question, dependent on how oil rents will be used in practice. For instance, if a 'spend all' strategy is used, not only will there be pressures on the local currency, with implications for imports and exports, but also the distribution of income may become more unequal between rural and urban households, as agriculture may be adversely affected by the appreciation in exchange rate. However, if a 'save all and invest interest only' strategy is used, there is less likelihood of Dutch disease effects. From this perspective, the current formula that guides how oil revenue should be used – a theme to which I return in the next few paragraphs – is likely to prevent Dutch disease effects. However, it is still early days, so it is more appropriate to regard Dutch disease effects as 'uncertain'.

Volatility is the second uncertainty. Actual oil revenue obtained on the world market may exceed or fall below expectations. It follows that there may be some periods of 'windfalls' and other periods of 'losses'. A priori, we would expect that these 'ups' and 'downs' would have implications for economic growth, as they can change revenue and expenditure plans in the economy. The empirical work of Van der Ploeg and Poelhekke (2007) shows that, in the long run, private volatility causes a negative relationship between natural resource rents and economic growth. The relationship is stronger when volatility is higher on the world market or when there is greater dependence on oil rents and less diversification in the economy.

To offset this aspect of the oil curse, the Petroleum Revenue Management Act (Act 815) has established a 'stabilisation fund' (see section 9 of the Act and also Breisinger *et al.*, 2010; Gatsi, 2010), to provide 'stability' in the face of fluctuating oil prices. On the one hand, if oil prices rise above budgetary projections, the 'excess' rents are invested in the fund. On the other hand, if prices fall such that revenues received are lower than projected, part of the fund may be used to finance the budget (for more on stabilisation funds, see Davis *et al.*, 2001) to lessen volatility.

While section 17 of Act 815 stipulates a range of 'not more than 70 per cent' of total oil revenue as the share to be used to support the national budget (called the Annual Budget Funding Amount, see section 17), the official government position is more definite: that 70 per cent of oil revenue will be expended to support the annual budget. The remaining 30 per cent is to be regarded as 100 per cent, which will then be split into a Heritage Fund[2] (30 per cent) and Stabilisation Fund (70 per cent) (Kwettey, 2010). Whether a stabilisation fund is sufficient to insure against price volatility, only time will tell. Such funds do not typically anticipate permanent volatility or long periods of price volatility (Davis *et al.*, 2001). Also, as the experience of superannuation funds in Australia shows (Frankel, 2004), the effectiveness of such funds is crucially dependent on where and how they are invested as well as the calibre of the fund managers and whether they are accountable to the populace. Indeed, were the funds invested in stocks and shares, they would be subject to the same price volatility they seek to stabilise.

Meanwhile, there has been considerable debate about how to spend the revenues that will go into the Annual Budget Fund. The concerns ranged from fiscal and monetary aspect or inflation to fear of embezzlement (Acosta and Heuty, 2009; Amoako-Tuffour and Ghanney, 2013). Policy makers and researchers are taking such concerns seriously. The government has initiated plans to establish an independent revenue management body and facilitate the establishment of an Oil and Gas Reporting Project. Todd Moss and Lauren Young (2009), researchers from the Center for Global Development in the USA, have recommended direct cash transfers to Ghanaians to increase their interest in how oil revenues are spent. Direct cash transfers, they argue, can make the state more accountable to the people because the state has to depend on ordinary people (with oil money) for taxation. Thus, rather than the state directly using revenue from the oil industry for the provision of public services, Moss and Young argue that placing the revenue in the hands of the citizenry – from current estimates, about $50 per adult per year – through direct cash transfer is the way to ensure transparent revenue management. Such recommendations deal with the social and political aspects of the concerns about oil rents, a topic to which I shall return in Chapter 9.

What about the economics of managing oil rents? According to Dagher *et al.* (2010), from the perspective of ensuring a healthy fiscal and monetary environment, the government should spend a small amount of the rents consistently over a long period of time. In this way, the benefits of the rents can be lasting and have sustained benefits. Moreover, it can help to check demand push inflation, arising from high short-term expenditure. In practice, the debate in Ghana has taken the form of concerns about whether to collateralise the oil rents. That is, the question of whether to borrow against the future oil rents. The position taken by the New Patriotic Party (NPP), which is the largest opposition party, and civil society organisations such as the Civil Society Platform on Oil and Gas is that borrowing against future oil rents is inimical to effective management of the revenues: it is a good principle in life that one does not eat one's chickens before they are hatched (*Public Agenda*, 2010c). This line has consistently been argued by the NPP to date.

The position of the NDC government, on the other hand, is to spend now, including spend future revenues by collateralising them (Ablakwa, 2010). The Parliament of Ghana has voted to allow collateralisation of the oil rents, all of which are supposed to be placed in the Holding Account, a temporary account, for subsequent disbursement (Section 2 (1–2) of the Petroleum Revenue Management Act, 2011, Act 815). According to PIAC (2012, p. 41), the Government of Ghana has entered into a loan arrangement with China totalling $3 billion which was collateralised against the Annual Budget Funding Amount.

The question about how effectively the rents should be expended remains contentious. Economic studies about this issue are lacking, save the work of Dagher *et al.* (2010). Dagher and his colleagues claim that if, on the one hand, there is little expenditure in the non-traded sector, the pressures on the real exchange rate arising from the realignment of factors of production from the traded to the non-traded is considerably diminished. From that perspective, if all expenditure is in the direction of the traded sector, the country risks less exchange rate pressures. It follows that if oil rents are expended on projects with high import content, the tendency for the real exchange rate to rise would be low. Similarly, less expenditure on non-tradable goods such as services is likely to exert lower pressure on wages and inflation in the services sector. On the other hand, high expenditure on traded goods in the real sector of the economy would benefit both the tradable and non-tradable sectors, following a Keynesian logic of economy-wide impact, resulting from public investment in the 'real economy' in areas such as road construction, building schools and hospitals (Keynes, 1936, 1973).

If the latter scenario is realised, Dagher and his colleagues (2010) argue that inflation may rise, suggesting the need for prudent monetary policies. Further, they suggest that, if the Bank of Ghana saves up a significant share of the rents from oil in foreign currency with the aim of controlling real appreciation in the exchange rate, it may have long-term problematic consequences for inflation. They contend that, while whichever policy is adopted has costs and benefits, overall, a combination of public investment and spending more on items with high import content may be useful to reduce the extent of the fiscal and monetary issues related to oil rents. The Annual Budget Fund Amount as of 30 September 2011 was $112 million (Ministry of Finance and Economic Planning, 2012). Details of how this amount was spent are contained in Table 3.4.

Road construction dominated in the expenditure pattern, suggesting possible employment generation potential (we shall return to this issue in Chapter 9). Also, the expenditure on 'oil and gas infrastructure' mainly went to the

Table 3.4 Allocation of Annual Budget Fund Amount (ABFA)

Item	Percentage of total ABFA
Expenditure and amortisation of loans for oil and gas infrastructure	11.90
Road infrastructure	79.82
Agriculture modernisation	7.83
Capacity building (including oil and gas)	0.45

Source: Ministry of Finance and Economic Planning, 2012.

establishment of the Ghana Gas Company – again an avenue for job creation. The budget statement clearly discloses details of the various expenditure heads and each of the expenditure items is selected on the Ghana Shared Growth and Development Agenda, one of the roadmaps for Ghana's economic development (PIAC, 2012). Although it is telling that capacity building receives very little attention and hence can lead to a situation where there are oil jobs but not sufficient local expertise to take them, it seems the only downside with respect to revenue expenditure which PIAC notes is that the budget does not capture expenditure for the fourth quarter of 2011. Even then, PIAC suggests that this setback is understandable because the data were not available at the time of preparing the budget and the Ministry of Finance and Economic Planning seems to have started taking steps towards rectifying the hitch (PIAC, 2012). Thus, there is no evidence of 'mismanagement' or inefficient management. Indeed, the first Petroleum Transparency Index published by the Institute of Economic Affairs, one of Ghana's leading independent think tanks (*Ghana News Agency*, 2012; Asafu-Adjaye, 2012, for the detailed analysis), gave the then Minister of Finance and Economic Planning a perfect score for making information on oil and gas revenues and expenditure readily accessible.

It has been six years since oil production begun and there is no overwhelming proof of corruption either of the media or of the public service, cronyism or widespread nepotism. Table 3.5 shows that Ghana's experiences with open discussion, safety and rule of law, participation and human rights, sustainable economic opportunity, and human development, as measured by two global and well known governance bodies, have remained positive and very strong both in Africa and the world.

Furthermore, the media has remained independent, as column three on Press Freedom shows. While there have been some swings in the rankings since 2005, the worst rank was attained outside the oil era (2005), while in 2009 – when oil had already been discovered and plans were underway to drill it – Ghana obtained the best rank in the last eight years. Between 2007 and 2013, Ghana's rank in the World Rankings of press freedom worsened by one point only, that is, from 29th to 30th. Finally, the perception of corruption – as measured by Transparency International[3] – has declined since oil was found (see Table 3.6).

Table 3.6 does not support the view that oil extraction has led to a systematic increase in corruption. Instead, in terms of both score and rank, the perception of

Table 3.5 Ranking of the Ghanaian state and its relationship to the media

Year	Mo-Ibrahim Governance Index Rank in Africa	Press Freedom Rank in the World
2005	8	66
2006	8	34
2007	8	29
2008	7	31
2009	7	27
2010	7	26
2011	7	41
2012	7	41
2013	7	30

Sources: Mo Ibrahim Foundation, 2012a, 2012b; Reporters Without Borders, various years.

Table 3.6 Ghana's Corruption Perception Index rank and score, 2006–2012

Year	Rank	Score: most corrupt (0), very clean (10)
2006	70	3.3
2007	69	3.7
2008	67	3.9
2009	69	3.9
2010	62	4.1
2011	69	3.9
2012	64	4.5*

Sources: Transparency International, 2006–2012.

Note
* This was originally stated in the 100s (45) but, for consistency, was divided by 10.

corruption substantially improved between 2007 when oil was found and 2012, the last year for which data are available. Viewed together with the other experiences of the state, it can be argued that the Ghanaian state does not systematically pursue its own interest, neither does it single mindedly pursue the public interest. Rather, it tends to do more by way of supporting particular class interests, although massive corruption – as has happened elsewhere – is not a major part of the Ghanaian experience.

Conclusion

The argument in this chapter is not that the resource curse and rentier theses are wrong, but that the evidence shows that they have to be greatly qualified and problematised. There is cause to be unsure about the political economy of oil, perhaps oscillating between optimism and pessimism according to how multiple and coterminous possibilities impact differently on various actors. While the Ghanaian state is gradually becoming a rentier, in the sense that oil rents are increasingly becoming a large share of national income or that it obtains substantial amount of revenue from the sector, there is no evidence of total neglect of human and social rights, oppression, and systematic and widespread corruption at this stage. Further, economic growth within the oil era has actually been rising. From these perspectives, the excitement about the discovery of oil must be tempered, but so must excessive fear of a curse.

As I write the conclusion for this chapter, my attention has been drawn to interesting claims by Dominik Kopiński and others (2013, p. 583) whose paper has just appeared in *African Affairs*. They reach similar conclusions to mine, namely that evidence of a stable political system, strong macroeconomic conditions, and vibrant civil society presence proves that a 'Nigeria scenario' is unlikely in Ghana. But, and this is an important but, they go as far as claiming that Ghana has an 'immune system' to resource curse. While, the evidence so far bears out the first part of their argument, claims of 'immunity' from resource-related challenges appear to be an exaggeration.

There is the need for more careful empirical analysis beyond evaluations of presence or absence of a curse or rentierism. Beyond claiming that 'local content' in terms of employment in the industry is high, there is the need to ascertain which types of jobs local people obtain, as well as to analyse the gender aspect of employment. A political economic way of seeing is more appropriate for exploring the implications of such questions. It may explore expectations and their construction and the role of capital in shaping the society, environment, and economy of oil communities. Of particular interest are questions of equity and how they are likely to be re-configured because of oil production. More careful and systematic analyses of the urban level economic, social, environmental, and institutional dimensions are required to complement the broad, national level account. It is to a consideration of these issues that the rest of the book turns, beginning with the early life of the oil city, the twin city of Sekondi-Takoradi.

Notes

1 The official website of the Electoral Commission of Ghana, www.ec.gov.gh/page.php?. page=376&section=45typ=1 (accessed on 21 May 2012), contains details of the registered parties. The details of the Progressive People's Party, registered in 2012, are yet to be included.
2 Fund set aside for use by 'future' generations (see section 10 (2a) of Act 815).
3 Transparency International, a global reputable organisation, analyses a wide array of surveys on perception of corruption in the public sector from which it compiles the Corruption Perception Index for countries on an annual basis.

Part II

From fishing settlements to oil city

Photo 3 The historic railway station building in Takoradi.

Photo 4 A cross-section of houses, cars, streets, and people in Takoradi.

Photo 5 Entrance to the Takoradi Oil Village gated estate.

Photo 6 Fishers' accommodation and meeting place in Sekondi.

Photo 7 A fisherman mending his net in Sekondi.

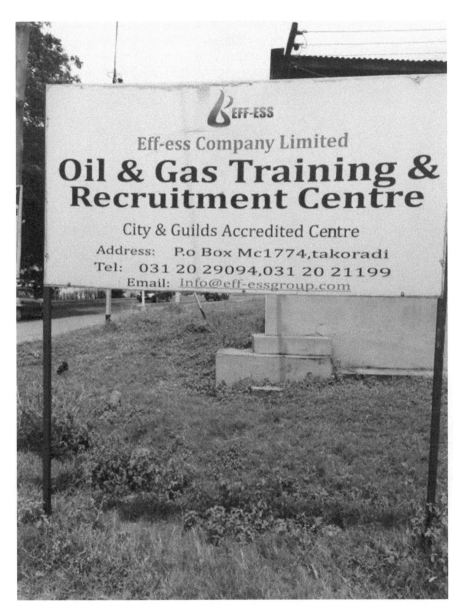

Photo 8 The signpost of an oil and gas training and recruitment centre in Sekondi-Takoradi.

Photo 9 Takoradi Oil Field Centre.

Photo 10 Ongoing construction activities, Takoradi.

Photo 11 Fort Orange in Sekondi.

4 Sekondi-Takoradi

The twin city and its history

Introduction

It is important to consider whether the future of Sekondi-Takoradi will be as bleak as resource curse theorists suggest or as rosy as the 'oil bonanza' optimists anticipate. One way to do so is to analyse how the city began its life and how it has fared to date. Existing studies on the city, looking at oil (see, for example, Boohene and Peprah, 2011; Yalley and Ofori-Darko, 2012), have not been historical. So, this chapter takes a step in that direction. It looks at the role of political and economic institutions in the origins and trajectory of Sekondi-Takoradi in the last 100 years; and examines how the city experienced periods of sudden change and fame in the past. The chapter is important because by trying to understand the present and reflect on the future through examining the past, it tries to provide insights for evaluating the claims of the resource curse or bonanza doctrine perspectives.

In doing so, it uses an 'institutional-analytical' method in the *longue durée* tradition which engages historical details and records and relates them to broader social explanations embedded in the roles of political economic institutions and phenomena that are interlocking and interdependent (Tuma, 1971). This approach entails critically assessing and synthesising micro and macro level qualitative and quantitative historical data. It simultaneously traverses and transcends the 'economy', the 'society', and the 'polity' in accounting for the 'economic' (Gay, 1930; Greif, 1998). Specifically, the method builds on (1) textual (reading original texts to tease out the intent of the authors); (2) contextual analysis (ideas by taking into account the mood of the period of time); (3) historical narrative (including critical synthesis of the stories in the past); and (4) rational reconstructions (such as re-reading old texts and making sense of them in a modern context) (Marcuzzo, 2008).

The chapter shows that the rise, growth, and trajectory of Sekondi-Takoradi to date, driven by its transport and trade subsectors, were shaped, constrained, and restrained by the interactions of institutions locally, nationally, and internationally in a process mediated by the agency of workers and activities of other social groups. From this perspective, it is argued that the expectations of the oil economy and its effects cannot be understood as an 'event'. Rather, oil dynamics ooze through a complex web of social relations and institutional interactions in an urban economy which is not only colonial but also colonising. These findings suggest that narratives that posit a determinist relationship between sudden change, such as a resource boom, and 'social disruption', 'resource curse', or 'rentier state' can be misleading. The ensuing five sections discuss, in turn, the pre-colony of the twin city and its early experiences with Europeans, the colonial, independence, and post-colonial periods, and the age of oil.

The pre-colony; and early encounter with foreigners

'Takoradi' and 'Sekondi' are indigenised forms of the original Prussian names, 'Taccarary', and 'Secundis'. They were the names of two fishing settlements in the Gold Coast.[1] The people of Sekondi-Takoradi were mainly Ahantas, a subgroup of the Akan people. While the origins of the Akan people, especially their purported connection to the ancient Kingdom of Ghana, constitutes a major source of debate among historians (Goody, 1968), it is well established that the Akans are the largest ethnic grouping in Ghana. The key subgroups are the Ashanti, Akim, Akwapim, Brong, Kwawu, Assin-Twifo, Wasa, Fante-Agona, Nzima, and Ahanta (Nukunya, 2011, p. 33). The Ahantas are mainly fishers and farmers and occupy the coastal strip in South-West Ghana, spanning Sekondi-Takoradi to Half-Assini in the mineral rich Western Region of Ghana (Agovi, 2008, pp. 47–61). 'Kuntum'[2] – derived from 'Kofi Kuntum', a god of the Ahantas – is the annual festival of the people (Sekondi-Takoradi City Council, 1963). Like elsewhere in Ghana, during this festival there is great merry-making often preceded by a period of solemn reflections and prayers.

The inhabitants of Sekondi-Takoradi have had a long association with Europeans. As early as the fourteenth century, they were trading with Europeans, and by the fifteenth century trade was said to be brisk in the area. The Prussians were one of the first foreigners to establish trade contact with the Ahantas and Sekondi-Takoradi in particular. They did so even much earlier than the Swedes who were trading there as early as 1650. Nevertheless, it was the Dutch who first settled in the city around 1640 during which time the French, Danes, Swedes, and English were in Takoradi. In 1644, the Dutch built Fort Orange ('Oranje' in Dutch) in Sekondi (Sekondi-Takoradi City Council, 1963; Dantzig, 1980; Owusu-Ansah and McFarland, 1995) and in 1659 they built Fort Witzen in Takoradi, although they abandoned it in 1818 (Owusu-Ansah and McFarland, 1995).

The forts were mostly built with the consent of the local people, but they were not always used for what they were ostensibly built for: storehouses for goods to be traded (e.g. guns and gun powder) and goods bought in the Gold Coast (e.g. gold and ivory), the defence of Europeans against other Europeans; and the marking of the presence or 'territories' of the builders. Instead, they were at various times used for 'illegitimate trade', that is, trade in slaves (Dantzig, 1980). My interviews at Fort Appollonia in neighbouring Nzemaland in the Western Region and at Cape Coast and Elmina Castles during fieldwork suggest that slaves, including some from Sekondi-Takoradi, were kept all over Ghana.

Between 1701 and 1704, some 2,328 people were shipped out of Ghana as slaves. The number rose to 10,198 between 1704 and 1707 (Ministry of Education, 1991, p. 211). A vivid description of the degrading and dehumanising conditions of slaves and how they were treated is given by the slave ship captain, John Newton as follows:

> The slaves were chained two and two together and laid in two rows one above the other, on each side of ship close to each other like books upon a shelf. I have known them so close that shelf would not easily contain one more. And I have known a white man sent down among the men to lay them in these rows to the greatest advantage, so that as little space as possible be lost. And every morning perhaps more instances than one, are found of the living and the dead, chained together.
>
> (cited in Ashun, 2004, p. 48)

Another slave ship captain Drake corroborates the account:

> On the eighth day (on the high seas), I took my round of the half deck holding a camphor bag in my teeth for the stench was hideous. The sick and the dying were chained together. I saw pregnant women giving birth to babies whilst chained to corpses, which our drunken overseers had not removed.
>
> (cited in Ashun, 2004, p. 48)

Stresses like these continued but the desire to obtain workers for plantation farms in the global centres of power continued to fuel the desire for more and more trade in slaves. Admittedly, the locals participated, sometimes gleefully (Ministry of Education, 1991). It is noteworthy, though, that the understanding of the locals regarding slaves was not shared by the Europeans. For the locals, slaves could marry into the royal family and they could become royals, if they worked hard enough and proved themselves competent and capable. Thus, local slaves enjoyed great social mobility in class terms. In contrast, the Europeans demeaned slaves, abused, and exploited them (Ashun, 2004) – never considering them to be in anyway of equal status with free people.

Through the work of humanists such as Granville Sharp and William Wilberforce, the slave trade was abolished in 1807 for British subjects and in the entire British empire by 1833. Other countries also abolished slave trade (e.g. the Netherlands (1814), Portugal (1815), and Spain (1817)). However, illegal trade in slaves continued until about the 1880s when it became inconspicuous (Ashun, 2004). Since then, the forts in general reverted to their past use for trade in 'legitimate' goods but also for residential purposes by Europeans.

Trade took the form of selling on either itinerant ships or establishing trade posts for companies in England to sell on the coasts. The merchants were keen to continue to profit from the Gold Coast and hence lobbied to have Britain use state power to ensure that their trade in the country was prosperous (Reynolds,

1975) and devoid of the recurrent threats by Ashanti such as those that happened in 1807, 1811, 1814–1816, 1823–1824, and 1826. George Maclean was appointed in 1830, among others, to govern British interest and to ensure peace between Ashanti and the coastal states, including Sekondi-Takoradi. During his time as governor (1830–1843), trade boomed and propelled some *individuals* to become active traders, a departure from the nature of trade before his time which was mainly chief-led. Under Maclean, strategic alliances between individuals and chiefs came to characterise trade in the Gold Coast. Such individuals became known as 'merchant princes' (Reynolds, 1975; Dumett, 1983) and their rise was part of a process of opening up trade to the masses – a process greatly enhanced by the development of steamship technology which made it possible for London to send more goods to the coast at short distance intervals.

The development of a credit system was a major factor which also made it possible for the rise of the merchant princes, but it was also responsible for their decline. Credit was given to the Gold Coast merchants, in particular, on usurious terms which often led to default, declining profits, or bankruptcy (Dumett, 1983). The competition between African and European merchants was, therefore, not always a matter of equanimity. Rather, the system worked against the Africans whose property often fell to their European lenders – by operation of Bankruptcy Law in 1858 (Reynolds, 1974). So, while between 1865 and 1900, there would have appeared to be some 200 merchant princes in all the coastal towns, the number dwindled quickly (Dumett, 1983). In contrast, the British companies extended their influence by becoming more visible on the coast.

British purported 'protection' of the coastal states, then, grew, although it failed to hold against the 1863 Asante attack. Also, imperial Britain purported to exchange forts on coastal lands with the Dutch without consulting the coastal people. This plan was fiercely opposed by the coastal states of Appollonia, Dixcove, and Sekondi; and the ally, but non-coastal state of Denkyera and Wassaw. Together, these developments drove the Fante people to found the Fante Confederation in 1868 (Laumann, 1993). Of these two reasons, the preponderance of the historical evidence seems to point to the defence of Akan states against Asante recurrent attacks as the more pressing concern for the confederation (Laumann, 1993). Not surprisingly, the British state itself showed no interest in relinquishing its economic and political interests in the Gold Coast. Rather, it took steps to expand it. Indeed, by a Convention signed in The Hague in 1871, the Dutch transferred all its rights of 'possession' to the British, in lieu of a fair price for Dutch commercial entities such as stores and the Island of Sumatra in Indonesia (Bennion, 1962). In 1872, the Dutch left the area and sold their belongings to the British (Administrators, 2012). In turn, the political division between 'Dutch Sekondi' and 'British Sekondi' collapsed officially.[3] Great Britain, looking at the substantial economic advantages of the Gold Coast (Reynolds, 1974, 1975), declared it a 'colony' in 1874.

Colonial encounter and administration

So, from 1874, the British officially became the coloniser of the Gold Coast. It seems trade continued to flourish and individual merchant princes came together to further drive trade. In 1882, the Gold Coast Native Concessions Purchasing Company was formed so were mining ventures formed in the 1890s. Also, in 1897, a local bank was opened. The coloniser strictly regulated the amount of loans given to Africans (Dumett, 1983), but some local merchants did well. In Sekondi-Takoradi, George Alfred 'Paa' Grant was on record to have been exporting some £2,000 worth of timber on an annual basis between 1903 and 1904 (Dumett, 1983). Purposeful public policy was put in place to ensure the growth of the private sector in Sekondi-Takoradi. For instance, there was a policy to reduce income tax payable by 25 per cent, if investors established business entities in the city (Owusu, 1998). Such policies encouraged the formation of the Bureau of African Industries in the twin city in 1934 (Owusu-Ansah and McFarland, 1995).

Trade in and between the twin city and others was important to the development of the city but even more important was the development of transport. We shall see in Chapter 9 that road transport is dominant in Sekondi-Takoradi today (Mahama, 2012), but it was peripheral to the development of the twin city. The first road network which was constructed in the Gold Coast was the Saltpond–Oda road. It was completed in 1895. There were only a few cars to make use of road networks around the time. Indeed, as of 1911, there were only 16 lorries and five cars in the Gold Coast, most of which were in Accra. However, the development of the country was rapid, so other networks such as the Nsawam–Cape Coast–Sekondi road were built in anticipation of an increase in the number of cars in the country (Ministry of Education, 1991), not that it was road network that powered the growth of Sekondi-Takoradi.

Rather, ports, harbours, and railways contributed more substantially to the rise and importance of Sekondi-Takoradi (Busia, 1950). Indeed, Albert Adu Boahen, Ghana's eminent historian, once described Takoradi as 'a product of the harbour' (Boahen, 2000, p. 105). Around 1903, the potential of Takoradi as a harbour city was discovered by a Polish Jew who planned to unleash that potential. He discussed his interest with the Chief of Takoradi, the custodian of the land, who agreed to lease the entire foreshore of Takoradi to the Polish man together with much of the land behind it. In exchange, the Polish Jew paid a monthly rent of £10. In addition, he donated a case of gin to the chief every month. Lefeber, as the Polish man was called, left the Gold Coast, but his successor continued to pay the rent and make the donation in the hope that a harbour could eventually be developed in Takoradi. In contrast, the colonial government focused its attention on trying to develop a harbour in neighbouring Sekondi (Correspondent, 1943).

Sekondi is older than Takoradi. The Sekondi Takoradi Metropolitan Assembly (2006) suggests that Sekondi had already emerged as a town as early as 1894. Although by 1912, it was being described as 'an upstart town' by one bureaucrat (Jeffries, 1978, p. 34), its growth and significance were rapid and impressive subsequently. A commercial precinct grew in Sekondi near the wharf, as a result of massive European investment in what eventually became known as European Quarter (Sekondi-Takoradi City Council, 1963) and exists to this day. Private investment aside, there was also public support of the city. Indeed, the first electricity installation in the country in 1919 was in Sekondi (Boahen, 2000). Takoradi was recognised as a 'town' much later, in 1926 (Busia, 1950) and had no reputation. One Correspondent (1943, p. 38) 'found Takoradi ... to be a small collection of dirty reed and thatch huts where the beach ended and the bush began'. Around 1913, the payment of the monthly rent was discontinued because the lessees no longer believed that Takoradi would ever become the home of a harbour (Correspondent, 1943). However, a harbour was built in Takoradi in 1928 as part of the ten-year plan of Sir Gordon Guggisberg, Governor of the Gold Coast at the time (Mendelson *et al.*, 2003). The Takoradi Harbour became West Africa's first artificial harbour (Hilling, 1975), complementing the role of the numerous ports in the Gold Coast but making Takoradi 'the gate-way to and a springboard for the economic and industrial development of the country' (Sekondi-Takoradi City Council, 1963, p. 23). This iconic harbour is widely regarded as the most expensive investment made by the British in its colonies in Africa in the 1920s (Plageman, 2013), so it occupies a significant place in the history of the city.

An important feature of the Takoradi Harbour was that it was served mainly by rail. That is, most of the goods sent to the port were carried by rail which was developed much earlier in Sekondi. While the Railway organisation started in 1901 (Ghana Railway Company Ltd [GRC], 2008), the first railway lines were built in 1898 in Sekondi to connect the gold mining towns of Tarkwa (1901) and Obuasi (1902), a distance of 41 and 123.5 miles respectively. Later, the lines were extended to Kumasi (1903). Other railway lines began from elsewhere, such as Accra, in 1905. Branch lines from Sekondi-Takoradi, such as the Tarkwa-Prestea line, were built in 1911. Sekondi was the headquarters of the railway organisation, as early as 1898 and, by 1915, Sekondi had grown considerably, mainly as a commercial hub where, lengthwise, some 490 miles of railways converged (Busia, 1950). In 1928, the Harbour in Takoradi was connected to Sekondi by a double line of track and, in 1934, the headquarters was moved to Takoradi when a new headquarters block was completed. By 1927, the Railway and Harbour were jointly administered, entirely by a headship of expatriate civil servants. The Railway and Harbours Administration was regarded as a statutory corporation for the first time on 1 July 1972 when the Railway and Ports Act, 1971 (Act 358) was passed (Busia, 1950; GRC, 2008).

The harbour and rail facilities had significant effects on the development of Sekondi-Takoradi. They attracted migrants to the city, especially from neighbouring coastal and other Akan communities. Most of these migrants became permanent residents and so Sekondi-Takoradi moved away from being a predominantly Ahanta city into being a mainly Akan city with Fanti, becoming the lingua franca. Nevertheless, they lived with other ethnic groupings (e.g. Ewe) and various groups from the Northern regions of Ghana harmoniously (Jeffries, 1975, 1978). A great number of the migrants contributed to the farming in the city and its surroundings. Indeed, the rail system greatly enhanced cocoa production (Jedwab and Moradi, 2011). Sekondi-Takoradi, Kumasi, and Accra collectively generated so much wealth and opportunities around rail transport that they were nicknamed, the 'Golden Triangle' (Owusu, 1998, p. 8). Between 1938 and 1939, exports from Takoradi increased by 186 per cent, while imports through the harbour soared by 133 per cent (White, 1955). Extension works were carried out at the Takoradi Harbour around the time and ended in 1956, significantly increasing its carrying capacity from two to three million tons per year (Hilling, 1975).

Harbour workers, according to Busia (1950), earned between £22 10s per annum and £1,000 per annum in 1947, working from 6(8) a.m. and finishing around 4(6) p.m. Outside these times, it was common practice, among workers in the country, to seek extra incomes elsewhere (Hart, 1973). Outside work in and around the harbour, there were many formal and informal markets in the city where the workers could earn extra income through various activities.

In spite of its economic prominence, Sekondi-Takoradi faced serious socio-economic problems, prominent among which were food insufficiency, housing inadequacy, and youth delinquency. Agriculture was a major employer in the urban economy, and fishers, hunters, and farmers were quite common in the city (Mendelson *et al.*, 2003). Yet the city typically imported its food (Busia, 1950), and the rate of new housing construction was significantly lower (163 per cent) than the rate of population growth (264 per cent) in 1970. Indeed, by 1984, the rate of housing development had declined drastically to 15.5 per cent (see Table 4.1).

Not only was housing inadequate numerically, it was also inadequate in qualitative terms too. Only about 17 per cent of the houses in 1950 had private toilets, and landlords were converting latrines and kitchens into spaces which they could rent out. Additionally, there was overcrowding in rooms and homes. More so, some houses lacked appropriate cooking facilities and, where traditional kitchens were available, they lacked storage facilities (Busia, 1950).

Table 4.1 Population and houses in Sekondi-Takoradi in 1948, 1970, and 1984

Year	Population	Houses
1948	44,130*	3,996**
1970	160,868***	10,507
1984	178,257	12,099

Sources: Busia (1950); CHF International (2010); Ghana Statistical Service (1989, 2005, 2012); The Consortium (2012).

Notes
* Excluding 807 non-African population.
** Refers to 1947.
*** City of Takoradi per se had 58,161 people, whereas Sekondi had 33,713 people, so the 160,868 figure referred to the conurbation (Pellow and Chazan, 1986, p. 97).

As the city expanded, juvenile delinquency increased too. Society was becoming complex and a large number of juveniles were regularly brought before the juvenile courts. Did these constitute evidence of anomie? Probably, but their causes could hardly be traced to just one sudden change. Long work hours, separation of parents, and breakdown of the extended family, the need for, but lack of, regulations to cater for new dynamics, and the creeping influence of capitalist production all contributed to delinquency. In turn, many children and youth began to fend for themselves by gambling and stealing, and sleeping in front of shops and in corridors in the market. Related problems of crime, prostitution, and corruption were also evident, as were problems of quarrelling, fighting, and unemployment (Busia, 1950). While the extent of these 'urban problems' has been questioned by some historians (e.g. Tait, 1951), their existence has not.

The Sekondi Town Council had the primary responsibility of managing the city. It was established in 1903 under the Town Council Ordinance No. 26. In 1943, the D.A. Sutherland committee was set up to consider the possibility of merging Takoradi and Sekondi. The Committee strongly recommended a merger in its report submitted in 1944. On 2 December 1946, Takoradi and Sekondi were merged politically and administratively as the Sekondi-Takoradi Town Council, with Sekondi as capital. Two other institutions assisted the Town Council to govern the city. The Native Authority, made up of chiefs and their councils nominated/elected based on familial, lineage, and clan lines; and the central government authority (Busia, 1950). The duties of these bodies differed and were often informed by ordinances, namely Native Authority Ordinance (1944), The Native Courts Ordinance (1944), and the Sekondi-Takoradi Town Council Ordinance (1945). The Native Authority exercised legislative, judicial, and spiritual responsibilities for which reason it was allowed to collect various taxes and rates. Significantly, it exercised its powers over only people of African descent. The Town Council was rather different. With a broader representation, every race was subject to its powers, which were mainly legislative, executive, and administrative. The council ensured that the people in the municipality were healthy by providing them sanitation services and health care facilities (Busia, 1950). It regulated business activities, building operations, and supplied other municipal services. The colonial Central Government had overwhelming powers in managing the municipality. It had the final say and decided whether a customarily nominated/elected and installed chief could, in fact, work in the Native Authority. This account, mainly from Busia (1950), also shows that, in 1954, three years before the country gained political independence from imperial Britain, the council was made a municipality because its population had grown substantially.

Protests, nationalism, and the march towards independence

Sekondi-Takoradi played a major role in the protest and nationalist activities that eventually led to political independence from Britain. One of the first acts of colonial dominance which was fiercely protested was the British attempt to expropriate Indigenous land through the Crown Lands Bill in 1894 and, later, the Lands Bill of 1897. By this bill, land in the Gold Coast would be controlled by the colonial state. As part of the proposed laws, all 'vacant land' was to be confiscated and administered as 'Crown land'. In principle, the colonial state wanted to lease Crown lands to entrepreneurs, be they British or natives but, in practice, to the richer class – the European bourgeoisie in the colony. Although that plan would have benefitted the native petty bourgeoisie too because they could afford to buy more land to enrich themselves, they feared the potential competition by European entrepreneurs who had more money (Howard, 1978).

The fear led to the formation of the Aborigines Rights Protection Society (ARPS) in 1898. The society was made up of the native merchant class – the petty bourgeoisie – native bureaucrats, the native educated elite, and the chiefs. Prominent leaders included John Mensah Sarbah and J.E. Kwegyir Aggrey, J.W. Sey, J.W. deGraft Johnson, and Rev. Attoh Ahuma (Quarcoopome, 1991). The ARPS petitioned the Queen, arguing that there was no such thing as 'vacant land' in the Gold Coast. All land had owners: occupied land was owned by an extended family or a community, whereas unoccupied land was managed by a chief or priest, as the case might be, for members of the community – dead, living, and unborn. It also argued that land often thought of as 'waste land' was, in fact, fallow land or land left to regain its lost nutrients (Howard, 1978). In addition, it waged a relentless local campaign against the bill. The challenge was successful and the coloniser withdrew the bill (Kimble, 1963). The ARPS subsequently transformed itself into a broader, nationalist movement concerned with anti-colonial matters (Asante, 1975). However, its victory song did not last for long as it suffered rancorous disagreements.

Sekondi-Takoradi legal practitioner and prominent member, J.E. Casely Hayford, therefore, broke off and founded the National Congress for British West Africa (NCBWA) in March 1920. As its name implies, NCBWA was much broader in its scope and its demands were more radical thereby. ARPS continued to be in existence, but its influence was eclipsed by the more outward looking NCBWA (Quarcoopome, 1991) which enjoyed considerable support from J.E. Casely Hayford's friends in Sekondi-Takoradi and elsewhere.

Within the Gold Coast, Sekondi-Takoradi and in particular its railway sector proved to be the vanguard of much political consciousness. In June 1918, for example, the first documented sit-down strike was organised by the Ghanaian railway workers, but it was in November 1928 that J.C.Vandyck, H.B. Cofie, W.A. Adottey, and H. Renner, F. H. Wood, S. W. Owiredu, and J.Eshun formally formed the first railway workers' union, dubbed Railway Association Committee, headquartered in a popular railway people's meeting 'joint', called Location in Sekondi (Jeffries, 1978, pp. 29–30). Since then, the railway workers recurrently staged demonstrations and strikes to press home their demands for improved working conditions for themselves and other workers. Most of the members of the Railway Workers' Union (RWU) lived in Sekondi-Takoradi, where RWU was headquartered. The workers could broadly be categorised into two, those in their 20s and 30s, and more elderly people. The former were typically more interested in ensuring a fair distribution of resources in society, while the latter usually focused more on internal politics of the union and the workplace (Haynes, 1991).

According to Busia (1950), railway workers constituted the single most populous group among formal sector workers' associations in 1947. They were more (3,276) than the total number of workers in the private formal sector (3,050) and more than workers in any government department, as shown in Table 4.2.

The railway workers had significant influence on political agitation and the nature of politics in the Gold Coast. Indeed, it is widely believed that their activities contributed substantially to the removal of the Progress Party government (1969–1972) from office (Haynes, 1991). A one-time leader of the RWU, Pobee Biney, who doubled as a leading member of the Convention People's Party (CPP) persuaded Kwame Nkrumah, leader of the CPP, to call for 'Positive Action' on 9 January 1950. By 'positive action', Pobee Biney meant a period during which all workers in the Gold Coast would boycott British shops and stay out of work. The motive was to force imperial Britain to grant the colony independence (Crisp, 1979; Ministry of Education, 1991). Nkrumah acceded to his request, catalysing other efforts to attain political independence.

The struggle for independence was closely related to the role of Takoradi in imperial Britain's effort to defeat the Vichy state – the breakaway state in France that supported Italy, Japan, and Germany, collectively known as the Axis Powers (Parker, 2005). In the lead up to the Second World War, around 1936, imperial Britain converted the small airport in Takoradi, established by (British) Imperial Airways, to military purposes (Jackson, 2006, p. 218). The Takoradi Airport was used as a strategic royal air base where British planes took off for North Africa and the Middle East to ward off Italian advances to the British Empire (Akyeampong, 2003). Between August 1940 and June 1943, more than 4,500 British aeroplanes were assembled in Takoradi (Jackson, 2006). Subsequently, Royal Air Force training schools were established in Takoradi. Also, the American army used the services of the airport as a 'portal to a major allied trans-African supply line' or, officially, as a major point in the West African Reinforcement Route (Jackson, 2006, p. 225).

The creation of the Royal West African Force was an additional level of support extracted by imperial Britain from its colonies in West Africa. That is, in addition to the use of the airport, a number of people in Takoradi were recruited

Table 4.2 Formal public sector workers, 1947

Department	Employees
Gold Coast Railways	3,276
Medical (Health) Dept. (Sekondi and Takoradi)	797
Public Works Dept. (Sekondi)	654
Takoradi Harbour	507

Source: Busia (1950).

into the Royal West African Frontier Force with a total enlistment of 65,000 Ghanaians (Israel, 1987; Jackson, 2006). While imperial Britain declared triumphantly that 'African soldiers beat the Italians.... They defended British West Africa from attack from Vichy territory ... and went to the Middle East as Pioneers and to the Far East to fight Japan' (Jackson, 2006, p. 171), the ex-service people began to feel it was time for their 'independence' and the liberation of their country (the Gold Coast) from the clutches of colonialism.

Many reasons account for this anti-colonial feeling. The ex-servicemen complained that they were paid less than the British officers. Also, they were mistreated, and discriminated against, although they felt they were as brave as the British in the army. According to one ex-serviceman, he had seen a British officer 'panting for breath and shouting for whisky' in the face of an attack, while a soldier from the Gold Coast had skilfully and bravely taken charge (Israel, 1987, p. 163). More so, they felt that infrastructure in some of the other colonies which they had seen during the war was better, in terms of both quantity and quality (Israel, 1987). In turn, the ex-soldiers started agitating for, and encouraging others to, fight for independence. As one veteran in Takoradi put it:

> In the array, the white people were not treating most of us well. They were kicking us, calling us 'you black monkey', and so forth. We didn't like the way they were treating us, especially when we went to Burma and all those places. So there we even decided to do something for ourselves when we came back.
>
> (Israel, 1987, p. 160)

Together with other local factors (some of which involved the leaders in Sekondi-Takoradi), national, regional, and international forces contributed to the attainment of political independence on 6 March 1957 (see, for a general account, de Smith, 1957). Contrary to Europeanised history of Ghana, a great number of these factors predated the Second World War and, apart from the explicit colonial-related reasons, there were political, economic, and social conditions that explain the roots of Ghanaian nationalism – revealed in pre-Second World War cross-ethnic protests against British imperialism (Boahen, 1964) – and subsequent attainment of independence.

Independence and after...

Independence brought great relief and joy to the now 'Ghanaians'. The American historian, N.A. Plageman (2008), documents the celebration with the local highlife music that attended the declaration of independence within Sekondi-Takoradi but also elsewhere. The lyrics of E.T. Mensah's famous *Ghana's Freedom Highlife* warrants full quotation here:

> Ghana, we now have freedom; Ghana, the land of freedom. Toils of the brave and the sweat of their labours; Toils of the brave which have brought

results. Kwame, the star of Ghana; Nkrumah the man of destiny. Toils of
the brave and the sweat of their labours; Toils of the brave which have.

(taken from Plageman, 2008, p. 220)

Yet, for Sekondi-Takoradi, activism had become a way of life. Indeed, after
independence, the railway workers resisted attempts to deradicalise and control
the union. As part of the post-independence resistance, a 17-day strike action
was called in September 1961 (Crisp, 1979). The union was characterised by
'labour aristocracy', a phenomenon of social differentiation among the class of
labour, but it kept a united front. Its activities further strengthened the identity of
Sekondi-Takoradi as a working class city (Jeffries, 1975) and a locale for
radicalism.

In June 1962, Sekondi-Takoradi was for the first time regarded as a 'city'
(STMA, 2010).[4] The city authorities at the time were explicitly left-wing.
Indeed, the President of the Sekondi-Takoradi City Council (Nana Baidoo
Bonsoe XIV), 'on the occasion of the elevation of Sekondi/Takoradi to City
status', offered 'hearty congratulations to the Chairman, Councillors and staff of
the Council on ... [the] elevation to City Status at ... [a] time of the country's
fight for economic emancipation and the *transformation of ... old inherited colonial regime to that of Socialism*' (Sekondi-Takoradi City Council, 1963, p. 7,
italics added). Kwame Nkrumah, the president of Ghana at the time, and the
person who promoted the local government to the rank of city government, was
a key advocate of socialism (Nkrumah, 1962) adapted to the African continent
or, what he called, 'scientific socialism' (J. Mohan, 1966, p. 220).

Nkrumah's industrialisation policies contributed substantially to the growth
of Sekondi-Takoradi. Indeed, the city did blossom, not only as the preferred
location for economic activity but also an attractive place for migrant labour
(Yeboah *et al.*, 2013). Notable among the industries of the time was the Pioneer
Tobacco Company about which, in 1957, Nkrumah observed:

I should like to draw attention to an excellent example of the kind of investment we appreciate in this country. I refer to Pioneer Tobacco Company.
Here is a young company working most efficiently among Africans, encouraging the growth of its raw materials ... this is an excellent case scenario for
private enterprise.

(cited in Owusu-Dabo *et al.*, 2009, p. 206)

Another aspect of the good days of Sekondi-Takoradi is the great seaman
culture, entailing overseas travels by seamen in the city and the stories of good
fortune that lay overseas. The expression, 'Sea Never Dries' captured the idea of
abundance of fortune hidden in the sea. The seamen or sailors were contracted to
work on ships either in Sekondi-Takoradi or natives of Sekondi-Takoradi travelled to Abidjan or other ports in West Africa or faraway in Athens, Greece to
obtain contracts to sail for a defined period of time. While sometimes uncertain,
work on ships usually made the seamen wealthy and this wealth was widely

spent on return to Sekondi-Takoradi amid tales of bravery exhibited by the seamen and stories of life overseas.[5] Thus, there was an established connection between Sekondi-Takoradi and the outside world, making it probably the first 'globalizing city', to borrow from Richard Grant (2009) who recently applied this descriptor to Accra. For, Chris Jeffries (1978, p. 21), Sekondi-Takoradi was a 'cosmopolitan city'.

The service sector in the city also grew and market activities increased. A prominent example was the Takoradi 'Market Circle', located in the central business district, and to this day regarded as 'the most organized market circle in Ghana'. It was built in 1960 and substantially expanded in the 1970s (*ENNIMIL*, 2011).

Both groups, however, benefitted immensely from the Sekondi-Takoradi Workers' College (STWC) established in October 1962. Although it was initially constructed as part of the Takoradi Hospital to serve the medical needs of workers, especially those working on the Harbour and nearby workers such as those in the railway service, it was turned into a learning facility. STWC made it possible for the railway workers – indeed all workers – to study for their GCE 'O' and 'A' level education. Indeed, around 1963, it had 1,200 worker-students studying various courses and tutored by 42 people who had graduated from secondary schools and others from state departments in Takoradi (Sekondi-Takoradi City Council, 1963, p. 33). In the late 1960s, the Sekondi-Takoradi Workers' College was passed on to the University of Ghana and, subsequently, structures such as the hospital were turned into a portfolio of staff bungalows, distant learning, and guest house facilities.[6]

It has been argued that the signs of decline in Sekondi-Takoradi started in 1962. Then, another harbour was built in Tema and for Francis Tawiah (2006) this was the beginning of the fall of the city. Others, however, saw the Tema Harbour as complementary. Indeed, it is widely taught in Ghana that the port in Takoradi became the resource export point obviously because it is located in the resource-rich Western Region, while the Tema Port took on the responsibility of manufactured imports.[7] However, just as the formation of a harbour in Takoradi eclipsed economic life in Sekondi, activities in Tema soon started taking a toll on Sekondi-Takoradi. Indeed, in 1972, Tema was taking 54.1 per cent of national port traffic and consistently doing better with national imports, taking some 69 per cent in 1972. Even in terms of efficiency, Tema seemed to be doing much better than Takoradi (Hilling, 1975). To Plageman (2013, p. 320), the new harbour essentially 'replaced' 'Takoradi as Ghana's most proficient port'. So, some people in Sekondi-Takoradi read conspiracy into Nkrumah's motive for building the Tema Harbour (Tawiah, 2006). However, to Tawiah, this view is unfounded, given the following reasons. First, Nkrumah was seeking general economic boom and was more intended to develop the entire country. Second, given that his own homeland, Nkroful, was in the Western Region, he could not have tried to undermine a facility that had brought so much prosperity to the entire region. Some of the early accounts of the building of the Tema Harbour (e.g. Kirchherr, 1968) did not impute any sinister motive to Nkrumah's actions so it is arguable that Nkrumah himself may not have intended his actions to have deleterious effects on Sekondi-Takoradi.

However, events and political choices after his overthrow would suggest that there were both overt and covert political reasons for the decline of Sekondi-Takoradi. The cumulative effect of lack of investment in the railway system, institution of government-sponsored parallel unions (the Ghana Trades Union Congress, in particular), the delinking of the administration of ports from the railways on 1 July 1977 by the Supreme Military Council using the Ghana Railway Corporation Decree (SMCD 95) and the Ghana Ports Authority Decree (SMCD 96), and the substantial expansion of road transport left the Railway Workers' Union weak (Haynes, 1991; GRC, 2008). Indeed, after the 1966 coup d'état in Ghana, culminating in the freedom of the Ghana Trades Union Congress (GTUC) from governmental control, the fulcrum around which workers organised shifted to the GTUC (see Goldsworthy, 1973) headquartered in Accra and, to this day, is the biggest labour union in Ghana. 'Unionists' and committees created by the Provisional National Defence Council (PNDC), a pretentious revolutionary-military establishment that was oriented to soliciting loyalty for the regime rather than the people infiltrated the remnant of the union (Haynes, 1991) and thereby weakened it.

These challenges for the twin city were part of a wider downturn in the economic and political life in the country. The period between 1966 and 1982 has been described by leading urban anthropologists Deborah Pellow and Naomi Chazan (1986) as 'a period of decline' in Ghana's political economy. That descriptor is not to suggest that there were no positive developments around the time. Indeed, Sekondi-Takoradi attracted many firms and, alone, contributed 21 per cent of the total wood processing firms in Ghana around the 1980s (Owusu, 1998). Also, in 1972, for example, Takoradi Harbour was the exit point of 80.4 per cent of total dry exports from the country (Hilling, 1975). Further, the Takoradi Police Station came into existence (1975) and its remit over major crime areas was extended in the 1980s (Owusu, 2010). Respectively, 70, 44, and 40 per cent of logs, sawn timber, and cocoa got transported to the harbour by rail in 1975. Port traffic was substantial. It provided 75 per cent of the cargo carried by the railway system. Around the 1970s, the unloading rate in Takoradi was the second highest in West Africa (Hilling, 1975). However, it was a period of decline in the sense that there were many military interventions, a severe drought, and a general global depression leading to dire consequences for urban and national life.

A sophisticated analysis of these problems has been offered by Ghanaian geographer Jacob Songsore (2011), but the Bretton Woods Institutions diagnosed these problems as a failure of the state-led model of development and proceeded to recommend, direct, and impose a market-led programme of social change. So, since 1983, economic efficiency or neoliberal turn in urban governance typified by a decentralisation model negotiated with the World Bank and characterised by accountability to market forces and little or no accountability to urban residents became the dominant public policy. The model provides little room for holding officials to account and numerous mechanisms for 'getting the prices right' in terms of downsizing, cut back of public funds, and outsourcing of local government and planning responsibilities to the market (G. Mohan, 1996; Yeboah and Obeng-Odoom, 2010). This model had deleterious consequences on all aspects of social, economic, environmental, and political life. On urban life, the pioneering work by Ian Yeboah (2000, 2003) has shown that the economic rationalist doctrine produced a particular type of urban form typical of capitalist urbanism: it led to the concentration of capital accumulation in the centres of wealth, namely Accra and Tema – and little 'development' elsewhere. So, while Tema had some natural advantages over Takoradi – as widely argued (Kirchherr, 1968), there were major political economic reasons that further made its rise detrimental to the once vibrant Sekondi-Takoradi.

According to the current Assistant Traffic Manager and one senior executive officer of the Railway Company, with whom I had a discussion in the historic railway station on the second day of January 2013, the straw that broke the camel's back was the (in)famous World Bank directed Rehabilitation Programme implemented in 1987 as part of Ghana's Economic Recovery Programme, Ghana's version of the Structural Adjustment Programme. The programme led to a renovation of trackwork and coaches, and the improvement of technical expertise *inter alia* through the secondment of expatriate staff to the service. The Government of India, in particular, offered financial and technical support to the Government of Ghana. Indeed, S.S. Nayyak, an Indian national, was appointed the General Manager of the service. While, as was the case with the Economic Recovery Programme as a whole, the World Bank directed a process of government withdrawal from supporting the service, the General Manager, S.S. Nayyak, was a staunch advocate of the withdrawal of government support claiming that the service would thereby be more effectively run.

Yet, this policy stance has proven to be detrimental. While the Railway organisation was issued with a Certification of Incorporation on 7 March 2001 with the intention of attracting more private sector interest, to this day neither investor nor politician has shown commitment to revamping the sector. In turn, the company has consistently seen a decline in almost all aspects – commodity transport, passenger service, and labour strength. We shall return to this issue in Chapter 9. For now, it will suffice to say that the cumulative effect of these political economic processes is that the city is governed by a highly deradicalised local government system, which is poorly funded, poorly resourced, and poorly motivated trying to manage what some, knowing the hey-day of Sekondi-Takoradi, often described as a dead city (Rémy, 1997). Housing conditions deteriorated, as did the state of general infrastructure (Owusu and Afutu-Kotey, 2010). Writers familiar with Sekondi in its hey-day could only euphemistically say that it had 'seen better days' (Rémy, 1997, p. 139). In Takoradi, the magnificent Railways Building stands, but only as a pale shadow of its buoyant days when its ancillary facilities and services were not decrepit. As a twin city, the city authorities declared the economy of Sekondi-Takoradi as lagging (STMA, 2012b).

Such is the story of this mighty city that people who know its past struggle to understand why Sekondi-Takoradi basically became a 'sleeping town'.

According to Joe Osae Addo (2013, p. 11), Chairman of ArchiAfrika, a professional group of architects and built environment practitioners, and a one time resident of Sekondi-Takoradi, 'Driving through [Sekondi] last week, my activist instincts kicked in and I wondered why this humane and well-planned town, of pedestrian friendly sidewalks and well-scaled buildings of handsome pedigree, had literally been left to decay'. A corroborator with whom he spoke said, 'No! Something must be done, Joe. I have lived through this decay and we cannot give up!!!' (cited in Addo, 2013, p. 12, emphasis in original). This enthusiasm has been buoyed by recent discovery of oil in the city, what politician-turned-scholar, Kwamena Ahwoi, is reported (Gyampo, 2011, p. 55) to have called an 'ATM Machine' for rapid urban, regional, and national transformation.

The age of oil

The year 2007 was a watershed in the history of the city. Oil was found in commercial quantity off the shores of Sekondi-Takoradi. Almost overnight, the city appears to have had a new lease of life. The Takoradi Airport, hitherto only a military base, now hosts two commercial aircraft companies flying people in and out of Takoradi almost throughout the week. Sometimes, the aircraft can make multiple trips,[8] an indication of the brisk business in Takoradi today.

Sekondi-Takoradi has been widely acclaimed as *the* 'oil city' because, apart from its proximity to the oil fields, it is the seat of key facilities and infrastructure such as the Takoradi Harbour and the Takoradi Airport. In turn, the oil tanker terminal of Ghana's oil fields is located in Takoradi, the twin of the Sekondi-Takoradi Metropolitan Assembly in the Western Region of Ghana. According to one researcher, 'Sekondi-Takoradi is the hub of the oil action' (Andrews, 2013, p. 67). So, this area is the obvious spatial focal point for this book, even though there are other settlements around the Jubilee field (or the settlements in the 'Jubilee Field Unit Area') such as Half Assini, Nkroful, Axim, Agona Nkwanta, Saltpond, Shama, Elmina, Cape Coast, Mumford, Apam, Winneba, Accra, and Tema (Tullow Ghana Ltd, 2009), Beyin, Etuobo, and Princess Town to which I make references precisely because of the reasons I outlined in Chapter 2.

Ultimately, however, it is Sekondi-Takoradi that is my main empirical referent. Any visitor to the premises of the Sekondi-Takoradi Metropolitan Assembly headquarters is greeted by a large and conspicuous insignia, *Nullis Secundi Sumus* or 'we are second to none'. Black gold was discovered off its shores in 2007.[9] By 2010, oil in commercial quantities was flowing and, since 2011, oil has been leaving its shores in pursuit of international currency.

This new phase of the city complements Sekondi-Takoradi's status as the third largest urban settlement in the country and the capital of the natural resources-rich Western Region of Ghana. However, at the same time, there has been a contagious crescendo of crisis narratives, drawing on the resource curse doctrine. The curse is so real, according to resource curse theorists (e.g. Sachs and Warner, 2001; Collier, 2006, 2008, 2009), and revisionists (e.g. Luong and Weinthal, 2010), that the chances of escape are slim and range from leaving oil unexplored, inviting international bodies or accepting their terms to scrutinise the use of the oil revenues, or entirely surrendering the management of oil resource to corporate capital as governments are so corrupt, so prone to corruption, or so inexperienced to deal with petro wealth in Africa.

Influenced by this doctrine, several media outlets in Ghana have published alarmist predictions with policy implications, ranging from 'Mr. President, Watch out! Oil money is coming, but corruption will follow' (Saminu, 2009), 'Oil will be hot potato for energy minister' (Public Agenda, 2009), 'Oil hot spot: Ghana must proceed with caution' (Hart, 2009); to 'Ghana: Why country should not bank on oil' (Osei, 2009). The Sekondi-Takoradi Metropolitan Assembly (2012b, p. 3, italics and emphasis in original) has recently expressed this dilemma as follows:

> Sekondi-Takoradi is at the cross-roads of the development agenda. An opportunity is beckoning for an otherwise slagging city to be transformed into a showpiece of an urban space ... in the mist of the oil and gas economy, the entire dynamics of the twin city could explode, but it could explode negatively into the most chaotic unplanned city ... or positively into that retrofitted dynamic, bustling *rare example* of a *liveable African* city!

Beyond the media and policy scare, the scholarly literature has been similarly influenced by the resource curse doctrine as analytical framework (see, for example, Moss and Young, 2009; Breisinger *et al.*, 2010; Dagher *et al.*, 2010; Gyimah-Boadi and Prempeh, 2012). Also, with a few respectable exceptions (e.g. Africa Progress Panel, 2013), regional bodies focusing on economic development in Africa have taken this view. Indeed, the *African Economic Outlook 2013*, published by the African Development Bank Group (African Development Bank *et al.*, 2013) looks from this perspective as does the United Nations Development Programme (UNDP) in its maiden *Western Region Human Development Report* (UNDP, 2013).

Finally, the general public seems to have been gripped by this economics of pessimism. A nationally representative survey of Ghanaian youth conducted by Friedrich-Ebert-Stiftung Ghana (2011) revealed widespread fear, reasonably inferred to be resource curse. Table 4.3 shows that 49.4 per cent of the sample agrees or strongly agrees that, while there is course to celebrate the macro-economic ramifications of oil exploration and exportation, there is surely going to be a disconnect between the national benefits and local impacts on livelihoods. Admittedly, recent national level research (e.g. Ayelazuno, 2013; Kopiński *et al.*, 2013) is beginning to question this widespread fear. Yet, these views – media, policy, scholarly, or public opinion – are mainly ideational, perceptual, or inapplicable to the urban question, as they are not based on systematic consideration of urban scale evidence. In turn, they are not helpful in evaluating the relevance of the resource curse doctrine at the urban level.

What they do show, nevertheless, is that the reach of the resource curse doctrine is wide. For that reason too, its predictions about curses ought to be more

Table 4.3 Attitudes to economic development: survey evidence for ten regions 'Our economic development will improve, but this will not reflect on the livelihood of the people'

Region	Ashanti	BA	Central	Eastern	GAR	North	UE	UW	Volta	Western	Overall totals
Strongly agree	25	12.20	14.05	17.71	16.68	16.13	19.35	19.49	11.72	23.69	18.27
Agree	32.14	27.65	33.64	37.06	30.92	29.22	35.48	28.93	25.59	31.01	31.13
Not sure	12.99	15.90	25.34	15.61	13.60	13.66	10.88	13.20	20.33	13.41	15.21
Disagree	21.94	35.41	19.58	24.70	25.29	29.41	26.20	28.93	25.35	19.33	25.05
Strongly disagree	7.89	8.90	7.37	4.89	13.49	11.38	8.06	9.43	16.98	12.54	10.32
Total respondents	1,039	528	434	429	941	527	248	159	418	574	5,297

Source: Friedrich-Ebert-Stiftung, Ghana, 2011.

Notes
64 Unanswered questionnaires.
BA – Brong Ahafo.
GAR – Greater Accra Region.
UE – Upper East.
UW – Upper West.

carefully studied, especially at the urban scale for which no prior substantive study has been conducted.

Conclusion

Some 100 years ago, West Africa's newest oil city was perceived as little more than a collection of streets of huts. It bore no visible signs of influence or affluence. A few bottles of gin and a small amount of pounds sterling were all that the 'owners of the city' obtained from foreign interests for the use of large portions of the city. However, with the effluxion time, the development of harbour and rail facilities, combined with worker activism, catapulted Sekondi-Takoradi to national prominence. From this perspective, the recent development of the oil industry and the attendant local, regional, and international interest seem to have a historical sibling. Sekondi-Takoradi has come full circle. It has seen this vibrancy before, although the post-2007 oil industry brings along distinctive challenges and prospects related to the development of land, labour, capital, and the state, locally, nationally, regionally, sub-regionally, and internationally. Social relations are likely to become ever more complex and new contours of adjustment and maladjustment may arise, and are feared to cause 'social dysfunction', 'resource curse', or 'rentier state' pressures. Institutions – formal, informal, implicit, and explicit, rules, processes, and customs – are predicted to be corrupted. However, these concerns ought to be verified empirically.

Notes

1 The name of modern Ghana before 6 March 1957.
2 The original name of the more recent 'Kundum'.
3 Nevertheless, to this day 'Boundary Road' that separated the two Sekondis remains and the *Omanhen* (Paramount Chief) of Essikado, regards that suburb as 'British Sekondi' in his book (Nana Nketsiah, 2013; interview with Mr Painstill, 2013).
4 The 'Sekondi-Takoradi Area' came into being in 1946 as a 'Town Council'. In 1954, it became a municipality and, in 1988, a metropolitan area, signalling the growing population of the area. Similarly, by 1988, the population had grown so significantly that the municipality was elevated to the rank of metropolitan area, operating under the Local Government Law, 1988, PNDCL 207 (Busia, 1950; Sekondi Takoradi Metropolitan Assembly, 2006).
5 I thank Albert Hagan, a former resident of Sekondi-Takoradi with extensive work experience in the harbour and with seamen of the city, and the historian, Dr Bob-Milliar for sharing these insights.
6 I thank the accountant at the Sekondi-Takoradi Workers' College, a worker at the College for some ten years, for calling my attention to these issues in a conversation on 30 January 2013.
7 I thank Ian Yeboah for this input.
8 Interview with Air Force Police/Leading Aircraft Man on 30 January 2013.
9 Since 2009, 16 new oil wells have been discovered, according to John Mahama, presidential candidate of the National Democratic Congress (NDC). This was said at the last Institute of Economic Affairs-organised Presidential Debate in Accra at the Banquet Hall of the State House on 21 November 2012.

5 Urban economic development in the age of oil

Introduction

The nature of urban economic development in Sekondi-Takoradi is poorly understood. On the one hand,

> stories abound in Takoradi about new investment proposals looming for the Western Region ranging from the design for a new city for 300,000 people ... as developed by Korean International Agency, through the rehabilitation of the railway link to Accra and Kumasi.
>
> (*Oil City Magazine*, 2012a, p. 24)

But others (e.g. Ackah-Baidoo, 2013) claim the oil industry is an enclave and produces no substantial changes on the urban economy. These sentiments are not based on systematic analysis of urban change in the twin city, so the empirical ramifications of oil exploration and production remain unclear.

This chapter attempts to fill that gap, drawing on economic data collected ethnographically – a well-established method used by Polly Hill (1966) and Keith Hart (1973), two of the leading economic anthropologists who have worked on Ghana. The data are made up of part interviews and part synthesis, and are discussed under four themes, namely employment, real estate development, transportation, and night life. This chapter argues that there has been a massive transformation in Sekondi-Takoradi's urban economy. These changes are driven jointly by the oil companies, the private sector generally, and the public sector. The chapter reveals that, while the available oil jobs on the rigs are clustered around the lower rungs and they are few compared with existing expectations, beneficiaries have been made better off, as have their family members some of whom have started small businesses. However, most of the jobs have gone to people who come from outside the city and outside the country. Further, obtaining a job is heavily influenced by ethnic and familial connections to onshore recruitment agencies. In principle, indigenous people can benefit from the boom in the property and construction sector, but the main beneficiaries here are chiefs, landlords, and wealthy indigenes. While the private financial sector has sought to capitalise on these changes to provide credit to poor taxi and commercial vehicle drivers, signs of default loom and days of overwhelming congestion can be predicted, looking at the trends in the financialisation of the transport sector. A common feature of the changing livelihoods and modern investments in the city is that, while some jobs have gone to women, most have gone to men: rig work, construction work, taxi driving. Besides, the gains in the winds of change have led to waves of eviction in the housing and land sectors.

Thus, overall, this chapter shows dark clouds existing side-by-side with the islands of economic privilege enjoyed by a few. The chapter is divided into four sections paralleling the key themes – employment, real estate development, urban transportation, and nightlife.

Employment

The ramifications of the oil and gas industry for livelihoods can be understood in terms of a supply chain composed of three key links, namely indirect services (e.g. catering and training), direct services (e.g. rig survey, field construction), and specialist services (e.g. subsea infrastructure and rig hire). It is the indirect services that are the most commonly sought after because they require the least skills, whereas the direct services are known to be difficult to find in the Ghanaian, local market. Specialist services such as well services can be found locally, but they are few and far between (Tullow, 2012). I found examples in Takoradi close to the Takoradi Oilfield Centre called Oilfield Machine Shop and another called Independent Oil Tools Ghana Limited. Both serve the oil industry in providing tool creation and maintenance services.

To date, livelihood changes and physical investment in Sekondi-Takoradi have been a function of at least three actors; the oil companies, the private sector generally, and the public sector. Oil companies work with the private sector, for example, to recruit labour and with the public sector, for example, to deliver education. So, the changes that are taking place in the city are a result of interdependencies to which we now turn.

An estimated 400 people have been employed by the oil companies to work on the rigs off the shore of the city. Most of these people are male; most have had international exposure often working in other African countries for oil companies; and most do menial jobs – cleaning, painting, scrapping, equipment packing, cooking, housekeeping, and cement handling. Significantly, most come from cities other than Sekondi-Takoradi. Employment on the rigs is an environment dominated by expatriates with long and established work experience. According to one informed observer, it should take another ten years for Ghanaians to be able to find jobs in the higher rungs of the oil rig employment sector (Ablo, 2012). Further, recruitment for the rigs is done largely along ethnic and familial lines. The oil companies outsource recruitment agencies which, depending on the ethnic background of the owner, will recruit people from that same ethnic group. To date, people mainly from outside Sekondi-Takoradi have dominated in employment on the rigs (Ablo, 2012). In particular people from the big cities such as the Accra–Tema city region are benefitting more from employment on the rigs. Further, medical examinations and aspects of the training to work on the rig are mainly conducted in Accra with some training done in Nigeria and related certification given by the West African Rescue Association and International Save Our Souls, two foreign companies – much to the displeasure of the Ghana Oil and Gas Service Providers Association (2010)[1] which argues that local medical bodies ought to be given such jobs.

Generally, the people who have obtained jobs have witnessed huge improvements in their economic circumstances. Indeed, there has been about 150 per cent increase in the average earnings of oil workers. The average income of workers before obtaining oil jobs is estimated at US$658.8, increasing to US$1,653 after obtaining the oil job, while the least income of US$55 jumped to US$306 (Ablo, 2012). This transformation has had multiplier effects, with employees reporting that they are better able to support their families some of whom have invested in small scale businesses/shops.

Overall, the employees report that they have enjoyed higher standards of living after obtaining employment on the rigs (Ablo, 2012). Nevertheless, rig workers are sometimes arbitrarily fired, denied end-of-contract benefit, subjected to unfair work practices such as screening out 'risky women' (those who are likely to become pregnant on the job), and have little room for career advancement, as revealed in Ablo's survey. Similar experiences are in the onshore, support services oil employment where, overall, professionals (e.g. estate officers) and non-skilled workers obtain better conditions of service. We know from the survey of Darkwah (2013) that there are many training and recruitment institutions, most of which are located outside the oil city. Indeed, some of them have overseas origins and all of them typically do not target indigenes of the oil city. One training institution – the main focus of Darkwah's study – is unique, but only because it targets people in rural areas. It is typical because its target population is not necessarily from the Western Region or the oil city. This inadvertent privileging of people in big cities, especially Accra–Tema city region, plagues even the Ghana Oil and Gas Service Providers Association which is headquartered in Accra and is, therefore, most active in its lobbying in the capital city.

The advantage obtained by non-Sekondi-Takoradi individuals over the indigenous population raises serious political economic questions of distributive justice. Yet, in what has been called 'that ten percent Imbroglio', the Parliament of Ghana refused to grant a request for setting aside 10 per cent of all oil revenues for the sole development and benefit of the Western Region in clear contravention of a promise of 'giving a pride of place as far as development is concerned' made by the then President of Ghana, John Mills (*Offshore Ghana*, 2011, p. 5). Indeed, a nationwide survey of the youth of the country revealed that a majority of them believe that greater access ought to be given to the people of the region (see Table 5.1).

The latest call for this issue to be corrected is contained in the *Western Region Development Report 2013* (UNDP, 2013, p. 13) in which we learn that 'there is a renewed determination among local citizens and their representatives to reduce the gap between the income the Region generates and the local benefits derived'.

Table 5.1 Respondents' opinions on whether the regions directly affected by oil and gas production should benefit more from oil revenues

Region	Ashanti	BA	Central	Eastern	GAR	North	UE	UW	Volta	Western	Overall totals
Strongly agree	25.83	15.6	24.09	21.6	20.89	20.41	18.95	15	21.19	52.09	24.72
Agree	32.55	52.89	38.63	39.54	35.1	33.94	35.88	40.0	40.23	2.0	38.49
Not sure	8.48	8.45	17.5	13.79	10.81	8.31	5.64	6.87	9.52	9.93	10.11
Disagree	20.68	13.34	12.95	16.32	23.01	20.41	27.01	25.0	20.23	8.36	18.40
Strongly disagree	7.43	4.69	6.81	8.73	10.18	11.90	12.50	13.12	8.80	3.65	8.25
Total respondents	1,049	532	440	435	943	529	248	160	420	574	5,330

Source: Friedrich-Ebert-Stiftung Ghana, 2011, p. 64.

Note
31 Unanswered questions.
BA – Brong Ahafo.
GAR – Greater Accra Region.
UE – Upper East.
UW – Upper West.

The concern is that 'it is both irregular and inequitable for resource-rich areas to experience a persistent drain of resources that systematically impoverishes local residents' (p. 12). What the report does not show, however, is that there are also intra-regional distributional tensions (see, for example, Andrews, 2013). But, of course, the major first cut is making the capital city of the region, the oil city, an icon through special treatment.

Yet, the Parliament of Ghana disagreed that the Western Region and for that matter its major twin city, Sekondi-Takoradi ought to be given preferential treatment. Indeed, the Parliament has been cited for not carefully putting first aspirations for local content in approving oil-related agreements. For instance, the well-known $3 billion Chinese loan offered to the Government of Ghana contains the provision that at least 60 per cent of all contracts related to the use of the loan must be awarded to Chinese companies (Odoi-Larbi, 2013). Studies of such Chinese loan facilities in Africa (e.g. Mohan, 2013) show that they invariably entail shifting resulting employment to Chinese labour, so there is a strong basis for critics to argue that most of the rewards of the oil industry will neither stay in the country nor with the local people.

In fairness to the oil companies, however, they have targeted some local communities to help them to improve their livelihoods. To better understand this programme, I interviewed one officer involved in implementing the Jubilee Livelihoods Enhancement and Enterprise Development (Jubilee LEED), a corporate social responsibility (CSR) valued at $600,000 (Agular, 2012) at his Windy Ridge office in Takoradi. I gathered that the Jubilee Partners, especially, are involved in community training in the six coastal districts of Shama (four communities), STMA (three communities), Ahanta West (six communities), Nzema East (four communities), Jomoro (five communities), and Ellebele (four communities). The people in the communities are trained in how to use their current funds or resources better (financial literacy), on how to develop new technologies (e.g. improved oven, better ways of rearing fingerlings), and training on how to diversify from current sources of earning (e.g. teaching soap making).

The partners are also implementing internship programmes for the youth to enhance their chances of seeking extra skills and to enable them to broaden their horizons. So far, 740 people have been trained since May/April 2012. There is a conscious attempt to make the CSR gender sensitive at 50 per cent each, although women and men tend to be attracted to different sources of activities. To ensure that the companies do not lose touch with the situation on the ground, community liaison officers have been employed. They link up the community with the trainers (also employed by the oil companies) and the oil companies. To date, it seems there is a wide perception of inclusion and inclusiveness in the programme. Fishers, farmers, and chiefs commonly partake in the training programme and the beneficiaries are given certificates of completion after finishing the training. According to the interviewee, the CSR is comprehensive with different teams involved in different aspects to ensure maximum returns to the communities.

To train manpower for the oil industry, Takoradi Polytechnic – a public institution – is preparing to organise short courses in oil and gas, sponsored and subsidised by the oil companies and the government. I interviewed the Dean of International Programmes on the nature of the programme, and the target group. Mechanical laboratories have been set up and the needed machines provided. The courses are meant for only 40 students, but 'a lot of people have expressed interest in the programmes', although the Dean said they keep no data on expressions of interest to enable me to analyse them. According to him, 'they [interested people] call and come here to make enquiries'. The short courses and longer (at least nine months) courses began in 2013 and cover four technical areas of electrical, mechanical, health and safety, instrument and process engineering activities at level 2 diploma. There will be ten students in each discipline. According to him, the training is vocational or technical, not managerial, and so is very likely to be employed at the lower rungs of the oil industry. Further, the students are mostly from around the country, with Sekondi-Takoradi not having any preference. I interviewed the Vice Dean too who took me round the laboratory to show me the computers, books, and other machines for the training.

While the Vice Dean acknowledged that the cost of training is high, he also mentioned the willingness of the oil companies, especially the Jubilee Partners, to support their activities and the training of manpower. Indeed, the $4.7 million ongoing Jubilee Technical Training Centre based at the Takoradi Polytechnic is funded by the partners in the Jubilee Fields to provide manpower for the oil and gas industry. The project, which started in November 2011, is about complete and four members of staff at Takoradi Polytechnic have been sponsored to do higher studies in the UK. Further, with the support of the Jubilee Partners, work on an Enterprise Development Centre (EDC) was completed in 2013. The EDC is going to train small and medium enterprises to benefit from the oil and gas industry (Mahama, 2013).

Finally, the stakeholders in the oil industry such as GNPC and the Jubilee Partners have supported private initiatives to train local manpower. Such is evidently the case of the Oil and Gas Training and Recruitment Centre in Takoradi where I interviewed the Business Development Officer and the Officer-in-Charge of training. The Centre, a private entity by two Ghanaians who set it up in 2011, is accredited by City and Guilds, UK. It is involved in vocational training of anybody who is literate, regardless of the level of education, for a six- to nine-month study period during which, depending on area of interest, instructors take students through four main courses, namely instrumentation, electrical, mechanical, and process monitoring, development, and implementation.

People who have no background in the engineering vocations may take nine months, while those with the requisite pre-knowledge can finish the course in six months. In future, it is expected that the centre will help in obtaining job placement for the graduates. For now, the emphasis has been on training. To date, about three batches of students have benefitted from the programme. The first batch, made up of some 16 students, entailed polytechnic graduates (14), one person with university degree and preparing to do a master's, and another person with a master's degree. The second batch of 13 students, included some people already working in various industries (about five) and the rest from polytechnics and universities. At the time of my interview, 15 students had already purchased forms to enrol as the third batch of students – although most had not returned them, so information on them was incomplete.

'Local manpower', from the perspective of the trainers, meant any Ghanaian willing to be trained. There is no dominance of Sekondi-Takoradi based people in the register of beneficiaries. Nevertheless, the total cost of training – in the order of Gh¢1,530 or some Gh¢2,030, depending on whether the person requires the optional extra three months of training in the basics of vocational engineering skills – seems to be lower than what pertains for similar training in Accra. The lower fee in Sekondi-Takoradi, according to my interviewees, is to help people from the city, generally said to be financially weaker than the people in Accra, to take advantage of the scheme. While this policy – if it can so be called – might indeed help people from Sekondi-Takoradi, as the records of the centre showed, most beneficiaries of the scheme are not locally based so they take advantage of a scheme supposedly designated to help the locals. This problem goes to the core of the *Local Content and Local Participation in Petroleum Activities Policy Framework* (Ministry of Energy, 2010) which defines 'Local content and participation … [as] the level of use of Ghanaian local expertise, goods and services, people, businesses and financing in oil and gas activities' (p. 2), without prioritising the needs of the people closest to the oil fields and most impacted by oil activities – for better or worse.

As of the time of my interview (28 February 2013), not one of the beneficiaries of the training had obtained 'oil jobs', although the trainers were optimistic that the trainees would become successful in their aspirations to become oil workers. This attitude of training providers, generally, helps to sustain or even create a false sense of employment Eldorado. Still, this evidence suggests that it will be difficult to meet the government's aspirations to 'achieve at least 90 percent local content and local participation in all aspects of oil and gas industry value chain within a decade [by 2020]' (Ministry of Energy, 2010, p. 5), let alone achieve the specifics:

> Management staff, at least fifty percent of the management staff are Ghanaians from the start of petroleum activities of the licensee and the percentage shall increase to at least eighty percent within five years after the start of the petroleum activities; (b) core technical staff, at least thirty per-cent of the technical staff are Ghanaians from the start of petroleum activities of the licensee and the percentage shall increase to at least eighty percent within five (5) years after the start of petroleum activities and ninety percent within ten (10) years; and (c) other staff, one hundred percent are Ghanaians.
>
> (pp. 8–9)

Further, there are strong grounds to be sceptical of the government's ability or commitment to achieve its promise that 'While Government will provide equal opportunities for all citizens of the Republic of Ghana, the participation of women in the oil and gas industry will be actively encouraged' (Ministry of Energy, 2010, p. 11).

The oil companies do more than employing people on rigs, supporting livelihoods, empowering communities, and supporting private sector and the state efforts at capacity training. Together with the policies of the government and private sector efforts, their activities have led to demonstrable impact on the real estate sector where through direct and indirect rentals, leasing, and purchasing, a major boom in the built environment has been registered in the twin city.

Real estate development

The arrival of the oil industry has triggered considerable activity in the real estate sector through, for example, renting of accommodation for their staff and renting of office space. The private sector is consequently positioning itself well to welcome investors. In turn, between 2007 and 2013, land value and prices have 'almost doubled' or 'increased significantly', according to my interviewees at the Lands Commission and Land Valuation Division in the metropolis. The cost rate used by valuers for contractor's test valuation increased between 88 and 107 per cent around the time, according to a well-known valuer in the city, interviewed for this study. Substantial numbers of new developments are springing up in the metropolis and selling even before they are completed.

Further, the real estate market has become increasingly dollarised and, consistent with a new exotic housing taste; new housing forms are being developed in the metropolis. For example, a gated and guarded housing community located between Ahanta West District and STMA, the 'Takoradi Oil Village', has been launched to target wealthy individuals and the oil workers in the Western Region. Rents range from US$1,000 to US$4,000, depending on the types of accommodation in the village, in a city where the median household income is $4.91 per day and most of the households (74 per cent) earn between $60 and $150 per month (CHF International, 2010).

Accompanying these developments is the emergence of a new class of consultants, specialising in real property transactions. Previously, estate agents were uncommon in the metropolis. Now, not only has there been an increase in the number of agents, their mode of operation and the economic class of people involved in the sector are transforming. New agents previously engaged in other economic activities have entered the real estate agency market. That is, agency has become an ad-on to other activities as it is becoming a viable economic activity.

Expatriates previously not involved in the estate agency business are increasingly becoming a key feature of real estate agency in the city. These companies are making considerable gains and are enjoying their job. Such is evidently the case of Dutchman Timo Voorn, founder and CEO of Takoradi Real Estate Ltd. Timo Voorn's formula for success is worth quoting:

> [w]e Dutch have a relaxed attitude which makes it easy to fit well in the oil city. Work hard, play hard and worry about the big issues but don't sweat the small stuff. Relax, work hard for your customers, listen to them, make them happy and show them you really care. It's a win–win you have to strive for. Cheers to oil city!
>
> (*Oil City Magazine*, 2012b, p. 40, emphasis in original)

Interestingly, they have not displaced the traditional ones. Rather, the traditional, moderately educated real estate agents have formed new alliances with the better resourced real estate agents to bring buyers and sellers together in the real estate market.

The activities of these real estate agents have substantial ramifications on the property market, not only through their traditional functions of bringing buyers and sellers together (Obeng-Odoom, 2011), but also through speculation for their own financial gain – given that they are more economically comfortable than traditional agents. The booming property market is also made up of 250 construction companies registered with the Registrar General's Department, in addition to many others that work informally, according to Owusuaa (2012). Like employment on the rigs, Owusaa found strong indications that most of the people in the construction sector are from outside Sekondi-Takoradi. Indeed, she found that one construction company based in Accra has projects in Sekondi-Takoradi. My own transect walks also revealed that one estate management firm in Accra has started working on projects in Sekondi-Takoradi. Also, like the real estate agency sector, the construction market is tiered. First, in terms of labour aristocracy in which workers in construction administration, often with better credentials, earn more than labourers and artisans. Also tiered are gender relations. Again, most of the jobs created in the construction sector are culturally defined as 'male jobs', educationally portrayed as such, or workplace norms regard them as such – driving most women away from them (Owusuaa, 2012). One training agency preparing people to take up onshore jobs in areas such as construction have in the last three years trained some 2,000 people, but only 20 per cent are female (Darkwah, 2013). From this evidence, it can be argued that the boom in the construction sector has mainly benefitted men. Given that the share of females in the population of the city is over 51 per cent, as we saw in Chapter 2, the jobs created in the construction sector are likely bypassing the majority of the population. However, generally, whether the workers in this industry are better educated or not so educated, they are casualised, and experience poor labour conditions and rights at different levels.

The real beneficiaries of the boom in the real estate market are landowners and house owners constituted by relatively few wealthy individual indigenes of the city, wealthy strangers who are investors, and speculators, and the chieftain class in the city. Each of these classes has a different set of property rights related to how land is owned in the city. Like other Akan societies in Ghana, chiefs are the custodians of the land. According to custom, native families can also obtain some land rights from the chiefs, usually called usufructuary interest in land. To the Omanhen of Esikado, whom I interviewed, land in the metropolis is not supposed to be sold. He writes in his unpublished book,

> Land is the foundation of ... the Akan's heritage, identity and spirituality. Land is the basis of Akan existence. The present does not even have the capacity to lease land. Land can NEVER be sold. As caretakers, the present is only in a position of trust.
>
> (Nketsiah, 2013, p. 290, emphasis in original)

However, in practice, chiefs are taking advantage of the booming real estate market to make a windfall – a finding confirmed by the eminent chief. Some lands have been 'sold' to non-natives or concentrated in the 'hands' of wealthy natives who, in turn, have converted the commons to private profit making estates. Often, the farmers tilling the land are said to have been compensated, although it is not clear how adequate or prompt is this payment. What is clear is that negotiations for such sales usually preclude common people or family members on the lower rungs of the family structure (a topic to which I return in Chapter 6).

Further, rental premises developed on common land are not meant for the poor. Similarly, existing rental facilities are no longer tolerant of poor tenants. Since the onset of the oil city fervour, if not frenzy, rental values in Sekondi-Takoradi have been arbitrarily varied upwards to push out poor sitting tenants and draw in richer tenants connected to the oil industry – dynamics of which I learnt at the Rent Control Department where, among others, I examined the department's *Complaints Book*.

It is in this book that complaints are recorded at the Rent Control Department in Sekondi-Takoradi. A typical list of entries since 2007 is made up mostly of recovery of possession. To recover possession, some landlords give reasons unrelated to oil opportunities, but investigations by the rent officers whom I interviewed, have established that repossessed properties are usually re-let to, or re-advertised for, higher income groups.

There is considerable evidence that previously existing rental units are being offered to wealthier classes, including the bourgeoning class of expatriate oil workers. While locals are sometimes welcome to the new market, there are conspicuous attempts to target expatriates. Locals who are drawn to or targeted are the wealthy class. During the launch of the Takoradi Oil Village, a ceremony I attended, most attendees were locals, but the very wealthy class as could be seen by the class of cars most of them drove, and their occupation.[2] The speech delivered during the launch suggested and the brochure distributed explicitly mentioned 'expatriate communities' (see p. 19 of *Takoradi Oil Village* brochure) being a key 'focus of the Takoradi Oil Village'. The interest in expatriates is now a common feature of the rental market in the city. Expat Solutions Ghana, another property broker, in the Beach Road area, has the tagline, '[f]or Expats'. My interview at Ussell (real estate agency) Services showed that it has prepared different brochures for locals and expatriates. Nevertheless, as the interviewee told me, 'we don't know the pockets of the locals, so they are welcome to check out the magazines meant for the expats'.

Like the residential real estate market, the leisure real estate market is being re-configured to suit the taste of the wealthy class. The sector has experienced a dramatic boost because of the oil boom, according to my interviewees at the Ghana Tourism Authority. The influx of high-income oil expatriates into the metropolis has led to attempts by hotels to try to meet their tastes. This attempt has been expressed in the form of new builds. For instance, Takoradi Raybow Hotel has demolished its six chalets to develop over 60 apartment rooms. The land for constructing new facilities may come from new buys, but also by buying out adjoining properties, or pulling down adjoining properties, and developing new ones from scratch in their place. An example of a provider using the latter approach to new development is Hill Crest. There are also those providers that are building new hotels in new areas. 'New hotels under construction' is how the *Oil City Magazine* advertises them (see, for example, *Oil City Magazine*, 2012a, 2012b, 2012c). Among these hotels are Western Atlantic Hotel that has about 224 rooms and Protea Select Hotel which has 136 rooms (*Oil City Magazine*, 2012b).

Not only are 'new' developments springing up, but also new standards have emerged. While before 2007 the most luxurious hotels were three-star rated, according to my informants at the Ghana Tourism Authority, currently there are five-star hotels such as Western Atlantic Hotel and, according to the *Oil City Magazine* (2012b), Protea Select, a four-star hotel, has also joined the real estate market. Other hotel providers have increased the number of rooms for their facilities, while some have turned their rooms into apartments, given that some oil workers prefer apartment housing in which to live for longer periods such as two years. Planters Lodge, for example, has increased its rooms from 19 to 45 since 2007.

These activities are creating new dynamics in terms of competition for 'oil clients' and the maintenance of old clientele. Oil clients, with oil money, are able to reserve huge hotel space for periods as long as two years. They do not stay continuously in the hotels, but they reserve the space in order to avoid the inconvenience of having to look for hotels in different areas anytime they are onshore or in the country. Hotel providers are, in turn, assured of a secure stream of income for a lengthy period of time. However, simultaneously, they face the dilemma of how to maintain their old loyal clientele. At the initial phase of the boom, some hoteliers made the decision to give out all their rooms to new 'oil clients', only to realise in the end that they had 'alienated' their old clients.

Now, the trend appears to be to keep a small number of rooms for old clients and reserve most of the rooms for the 'oil clients'. In turn, occupancy rates are artificially kept at around 90 and 95 per cent, although 100 per cent is possible in this present oil age. While in the early oil days, demand was far in excess of supply such that hotel providers tried to unilaterally review hotel rates upwards, as of 2012/2013, the oil companies have begun to carefully search the real estate market and look for competitive rates. Some clients are prepared to look outside the STMA area (e.g. in Tarkwa) for quality services and competitive rates.

Investors have responded to the demand-side dynamics of the leisure or hospitality real estate market. Currently, there are about 50 ongoing projects registered with the Ghana Tourism Authority, ranging from budget accommodation and one-star hotel, to guest houses (Ghana Tourism Authority, 2012). Since 2007, there have been important changes in the hospitality industry across the Western Region, as shown in Table 5.2.

Apart from first registration, a significant number of licences have been issued since the onset of oil production and development. In 2011, for example, 201 licences were issued, as shown in Table 5.3.

The number of existing units licensed is quite substantial, suggesting awareness that international clients are more likely to prefer formally registered hotels. Also noteworthy is that the Ghana Tourism Authority derives substantial revenues by performing its licensing activities.

Urban transportation

Like direct employment and property development, the urban transport sector in the metropolis is undergoing important changes. According to the city authorities, the number of cars, especially commercial ones, in STMA has increased to take advantage of the increase in the number of people in the metropolis.

Table 5.2 Western Region project registration, 2007–2012

Year	Restaurant	Informal sector	Hotel	Guest house	Home lodge	Night club
2007	4	13	22	6	2	1
2008	4	46	7	10	–	–
2009	1	14	17	–	–	–
2010	5	6	16	–	–	–
2011	12	–	27	8	–	–

Source: Ghana Tourist Authority, Western Region, 2012.

Table 5.3 Licences issued for various uses in the Western Region, 2011

Facility	Licences issued	Outstanding licences	New units licensed	Existing units licensed
Hotel	162	11	18	144
Restaurant	21	–	6	15
Home lodge	6	–	–	6
Hostel	1	–	–	1
Travel and tours	8	–	2	6
Night club	1	–	–	1

Source: compiled from Ghana Tourism Authority, 2012.

However, they had no numeric evidence to back up this claim. So, I followed up to the Driver and Vehicle Licensing Authority (DVLA) in Takoradi where I discussed registration trends with a Senior Executive Officer (SEO) with whose help I obtained some DVLA Monthly Return Files featuring data on vehicle registration in Takoradi. Apparently, the Authority only started disaggregating data on cars registered in 2012. While previously distinctions were made, for example, between 'motor vehicle up to 2000cc' and 'motor vehicle above 2000cc', 'buses and coaches' appeared as one category. So, it was difficult to determine which ones were put to commercial use to enable an analysis of trends in the registration of commercial vehicles.

I tried to extract data on trends in 2012 registration, but this dataset was incomplete. From the Monthly Return Files (page enclosed in the memo, DVLA. 12/ADM. 2012/1), it seems there were 4,532 newly registered cars in 2011 about 42.9 per cent of which were commercial vehicles. It is very likely that more cars were registered in 2012 and that the proportion of commercial cars increased between 2011 and 2012.

The evidence in the file, gives some indication that more cars, especially commercial ones, were registered between 2011 and 2012. Whether the increase in number and proportion of commercial vehicles can be ascribed to oil is hard to say, but the SEO said he would agree with the planners' observation that the number of commercial cars in the metropolis has increased, due to greater opportunities made possible by the oil industry in STMA. It is noteworthy, though, that depending on DVLA data can be misleading, not only because they are not collected in a consistent way but also because some vehicle owners would rather register their vehicles in Accra, owing to a mindset that everything Accra is superior or that Accra has better quality roads so a car with an Accra number plate may be deemed to be in better condition by prospective buyers who tend to think that cars with Accra number plates are mainly used in Accra.[3]

For all these reasons, I decided to hold informal discussions with some taxi and *trotro* or commercial minibus drivers on whose vehicles I travelled within the city. They confirmed this view but, even more importantly, I found stronger evidence of oil-induced transport-related investment. There are new drivers emerging in the metropolis who obtain car loans from the banks that are filing into the city to serve the new oil economy. My car journeys and transect walks around the city revealed signage of banks and other financial institutions that have advanced loans to people to purchase commercial vehicles. Such institutions include Ankobra West Rural Bank, Multi Credit, Lower Prah Rural Bank, and Unique Trust Bank. Of all these, Ahantaman Rural Bank appears to be one of the most popular. So, I visited the head office to discuss their scheme with the Head of Credit and one credit officer. Apparently, the bank operates its scheme along the microcredit principle. A group of five taxi drivers are required to form a team with a constitution. All the members of the team must have obtained licence for at least one year. The group acts as a whole in pressuring its members to pay up. Members make daily repayments for 24 months. The *Ahenkafo* (taxi) loan scheme was introduced in 2009. It is mainly for people interested in owning their car as taxi drivers, but civil servants interested in becoming commercial car owners may also be considered. A 10 per cent deposit is required and 5 per cent of the approved amount is paid as commitment fee used to take care of processing by the bank. Recipients pay a 25 per cent interest on the loan. According to the officers with whom I discussed the scheme, default rate at this stage is very low, around 5 per cent, so the success rate is around 95 per cent, 4 points below the bank's expectations.

It seems this new financialisation of the transport sector is good business. In turn, some private individuals have also become creditors. These creditors have purchased a fleet of cars for 'work and pay' purposes. So far, two of them are the talk of town, namely Angelica Travel and Tours and Ayax. More so, the government's free taxi scheme by which the government offers loans for the purpose of purchasing taxis, have helped poor drivers to obtain their 'own' cars.

There are other mediums through which the number of cars in the transport sector has increased. A key one is the growth in the car rental industry. My discussion with the Regional Officer of the Ghana Tourism Authority revealed that, while before the oil boom, there were only three companies involved in renting out cars, now there are some 12 companies registered and 13 licensed to deal in car rentals. Further, the oil companies are tending to support the development of road infrastructure. A case in point is the Shippers' Council Roundabout which has been renovated by PW Ghana Ltd, at the expense of the Jubilee Partners (*Oil City Magazine*, 2012a). The Government of Ghana also invests considerably in road infrastructure, using oil revenues. In 2011, for example, 76.5 per cent of the annual budgetary fund was invested in road infrastructure across the country (PIAC, 2012a). I shall return to the transport issue in the penultimate chapter. For now, it will suffice to say that the cumulative effect of all these transport investments is an increase in the number of cars in the metropolis, increasing congestion, but with it a feeling that the city is becoming modern.

Changes in night life: contested representations

A final conspicuous livelihood-related 'change' is night life in the city. Even at its origins, Sekondi-Takoradi had several pubs, clubs, and hotels. Some of these were The Sphynx and The Mainland Social Centre, both in Takoradi; and The Empire Night Club in Sekondi (Sekondi-Takoradi City Council, 1963). Currently, the old Zenith area, and Paradise (currently being refurbished) are key places for sex work. However, the most dramatic change, by far, is the establishment of Vienna City club, as distinct from Vienna City beach in Chapel Hill. This area is close to Akromah Plaza Hotel, the Shippers' Council Roundabout, and Spike's Corner. Only a short distance away are the Police Headquarters and the Bethel Methodist Church in Takoradi.

Vienna City is located at the Beach Road Roundabout. It is near the luxurious Akromah Plaza Hotel and Captain Hook Hotel in Takoradi, the twin of Sekondi. Vienna, as it is popularly called by taxi drivers, is open most of the time to its numerous customers. From 11 a.m. to 4 a.m., however, is the official time to visit this city. It has a casino, poker machines, and pool tables, among other electronic entertainment devices that can be seen in clubs in Australia, the USA, the Netherlands, the UK, and other technologically advanced countries. In turn, it is a magnet for many expatriates and people with a taste for 'overseas' experiences in a local setting.

The atmosphere in Vienna City can be, and often is, electrifying. A range of music genres blasts out from the loud, gigantic, and hi-tech speakers. From afar, one can have a foretaste of what lies in Vienna City. Right from the large car park, even beyond it, to the main club premises – indeed the entire Vienna City – one can see and feel the frenzy in the atmosphere.

In the midst of these activities, migrants in sex work, or the 'sisters', as some taxi drivers call them, are busily working: negotiating, persuading, or inviting ... sometimes through their great dance moves and good looks in a theatrically impressive setting. To one observer commenting generally about women he had seen in Sekondi-Takoradi, they can be very 'seductive' and their presence, 'an embodiment of the great beauty of its people and culture' (Addo, 2013, p. 11). So, anyone who has read 'City, culture and happiness' written by the happiness economist, Frey Bruno (2008) will be led to the conclusion that Vienna must be a very happy city and that migration scholars are indeed right: migration can offer sex workers a route out of poverty.

However, this would be a hasty conclusion. A great danger looms for the sex workers in Vienna City: their 'right to the city' (see Harvey, 2008, for a discussion of the concept) is under threat because most of the comments on the sex industry and the migrants working therein have been decidedly negative. Vienna City has been characterised as a hotbed of moral decadence of Ghanaian children, an avenue where people are indecently dressed, a spot for violent crime, to suggestions about the blighting effect of the industry on landed property (see, for example, Boyefio, 2012; Kokutse, 2012; Essien, 2011; Ghana News Agency, 2012, 2013). There is said to be a 'dark side' of the sex industry. In one article, titled, 'Child prostitution booming at Sekondi...', the *Ghana News Agency* (2013), observes that 'Child prostitution is flourishing at Sekondi landing beaches and other suburbs of the Sekondi-Takoradi Metropolis in the Western Region'. Commenting on prostitution in Sekondi-Takoradi, one ex-writer with *Offshore Ghana Magazine* reports the following encounter:

> an old gold miner ... expressed disgust at its [prostitution] ascendency and the careless and audacious manner with which current prostitutes now carry out their job. To him the practice has taken on a different dimension where even primary school children are being drawn into it.
>
> (Boyefio, 2012, n.p.)

According to the *Daily Graphic*,

> It is still too early to assess the full impact of oil and gas on the Oil City. However, there are few adverse impacts ... the increase in sex trade,

incidences of armed robbery and influx of young men from various regions in Ghana and also from neighbouring countries; crowding out the job avenues.

(Asamoah, 2013)

Admittedly, such sentiments are expressed generally about prostitution in the city. Take the example of Sam Mark Essien of the *Daily Guide Newspaper* commenting on oil experiences in Sekondi-Takoradi with the caption, 'Sex for sale: The rise of child prostitution in Takoradi'. He notes,

> Clad in a transparent white T-shirt covering only half of her buttocks, the little girl's breasts were exposed. The black high-heeled shoes she wore matched her Rasta hairdo.... Girls like Araba who are in the age bracket of 12 and 14 years are into prostitution all over the oil city of Takoradi. They look for cash from their clients, some of whom are high-profile persons.

He continues:

> The fee used to be GH¢5 per one bout of sex, but has now soared to GH¢10, with respect to the senior sisters. The child prostitutes have had their fee also revised from GH¢2 to GH¢5 in January 2011 when Ghana started pouring oil from its Jubilee Field.

(Essien, 2011, n.p.)

These claims require more careful property rights analysis based on data related to the specific depictions. To do so, my methods of collecting data were unobtrusive, non-participant observation over the three-month period during which I stayed in Sekondi-Takoradi. By this approach, I mean I visited the area regularly, carried out transect walks around the area taking photos sometimes, and once sat by a male sedentary mobile phone credit seller to catch a view of what was happening in the area and to take notes. I must say that my way of taking notes may differ from how an anthropologist would have done it. I am a land economist, but land economists too are trained to take notes (Obeng-Odoom and Ameyaw, 2011). Elsewhere, anthropologists have been generous in accepting my version of ethnography as in when they published my work in *Urban Anthropology* on urban informality. In any case, even people not skilled in taking field notes of any kind have been found to be capable of capturing crucial anthropological observations (Tjora, 2006). While there continues to be great debate on note-taking in anthropology, for pragmatic purposes it is one's knowledge of the existing research problem that largely determines what to observe and write (Wolfinger, 2006). I was clear on my aim for the research, so while I kept an open mind, the philosophical and epistemic issues about what to write did not become a barrier to my research.

I observed and took notes between 9 and 11 p.m., and walked in and drove around the area at various times of the day. Once, I got to listen to a detailed and lengthy conversation between a self-identified sex worker and her friend, a male trader in pay-as-you-go phone units/credit seller (often called 'phone card seller' in Ghana). I also visited the parking and seating areas in front of the clubs twice, and went into the club once. Further, I engaged taxi drivers who drive the sex workers around in the city as they go to see clients in long and repeated conversations about what they have learnt. One taxi driver, Long Story – so called because of his interest in telling his passengers his life story and personal experiences, and my personal taxi driver – was particularly informative in understanding this aspect of change and continuity in the city.

I would usually type my notes on my return to my accommodation where I would reflect on my notes, think through them and analyse them in the light of the depictions of the sex industry in Sekondi-Takoradi. On many occasions, I would raise some of my formulations and analysis with taxi drivers, with friends, family, and professionals in the city. Finally, I related historical and contemporary research conducted by others to what I observed and experienced.

For instance, the mode of operation of the sex trade in Ghana has previously been reported by Asamoah-Adu *et al.* (2001) to be of two kinds, namely the *seaters* and *roamers*. The former is made up of older women, around a median age of 37 years, often divorced or widowed who operate from their homes and they tend to have dedicated clients and have been in the business for a longer time. *Roamers*, in contrast, are younger, have been around for only a short time, and rarely make their clients use condoms. Finally, they tend to hang around hotels.

What is happening in Vienna City, however, does not neatly fit into these categories. There are aspects that are like seaters and others that are similar to roamers. Almost all the people involved are young and 'newcomers', obviously because this is a new settlement but, as we shall see, not all are children. The quasi-seaters are those who have their own place and can take a client 'home'. They hang around the roundabout or go into the club in the area to solicit or entice clients. Previous studies (e.g. Asamoah-Adu *et al.*, 2001) have assumed that these people are in full-time occupation, but what is happening in Vienna City is more about part-time work, with people working regularly outside the sex industry or studying and often or on occasion trying to make a living as escorts. Thus, the dynamics at Vienna City resemble what Keith Hart (1973) described some four decades ago in his seminal work on the urban informal economy of Ghana. It is part-time work, so sex workers have preference for expatriates or, in the words of Hart (1973, p. 76), 'bourgeois black men' often perceived as richer than the common local men.

Sex work is lucrative, especially as a part-time job. The mode of payment for sex services differs greatly. Some sex workers invite taxi drivers to offer them a free ride home in exchange for sex. Others take cash. The amount demanded is always negotiable, and contingent on clients' perceived social class and race. Other determinants are duration, and the number of 'rounds' spent with the worker. Wealthier clients can offer between Gh¢50 and Gh¢100 for different services, and there seem to be no differences in the number of rounds taxi drivers travelling different distances are offered. For all these reasons, it is difficult to answer the question, 'how much sex workers in Vienna City charge for their services', although media accounts provide categorical figures which are not always consistent with the everyday experiences of sex workers.

Yet, the sex workers seem to find the work lucrative. Indeed, they follow economic cycles to cut their losses, so now that the city has gained a new lease of economic life in the form of the oil operations, it has attracted many sex workers into the metropolis, as have investors. While I could not obtain any numeric evidence on the precise amount of money made per night, per week, or per month, I learnt that the rewards are great and help to complement other sources of making a living in the oil city.

In sum, Vienna City has been characterised as a hotbed of exploitation, moral decadence of Ghanaian children, an avenue where people are indecently dressed, a spot for violent crime, to suggestions about the blighting effect of the industry on landed property. First, some sex work can rightfully be regarded as exploitation of women or prostitution. That will be the case, for example, in scenarios where pimps and strip clubs extract a large share of the revenues sex workers in their 'employment' make such that the women actually might lose out in the end. Or, where women are forced into sex work by a sex dealer (for other examples, see Jeffreys, 2003).

It is doubtful, however, that the nature of sex work in Sekondi-Takoradi, especially the recent surge, is analogous to these examples – given their modus operandi. As roamers, they manage their own schedules and as quasi-seaters they do not usually work for an exploitative sex merchant. While it can be argued that they have been *forced* by difficult economic circumstances into sex work, this will not be the same as making the case of exploitation. Besides, that view will be counted by those feminists (see examples in Jeffreys, 2003) who argue that sex work ought to be regarded as legitimate work and hence the stresses therein ought to be ameliorated or managed as in other occupations; rather than regard them as exceptionally problematic. The evidence provides a strong basis to draw a distinction between 'sex work' and 'prostitution'. Both refer to exchanging sex for money, of course, but a sex worker exercises greater agency. Prostitutes, on the other hand, are usually exploited by others such as pimps (Vearey, 2013). The ladies in Vienna City are mostly *sex workers*; not prostitutes.

The second posited dark side of sex work in Sekondi-Takoradi is the wide and growing view that it is associated with crime increase. However, this view seems to be only a matter of perception. The metropolitan police force has not made any public statement to this effect. Indeed, the red light district is close to the Takoradi Police Headquarters which may actually explain the low crime rate in the area. Generally, however, criminologists (e.g. Appiahene-Gyamfi, 2003) have recorded very low levels of police-reported sex-related crimes in Ghana. One reason is that, while the police want to enforce the law, most think the workers need to survive their harsh economic reality. Sometimes, too, arrests are prevented by sex workers bribing the police (Bindman and Doezeman, 1997).

In the case of Vienna City, living around the area for three months continuously and paying visits to the area from time to time, I did not observe or hear any reported crimes in the area. But, of course, it is not enough to base my conclusions solely on what I heard or observed, so with the help of my research assistant, a Police Constable at the Takoradi Police Station located in the same neighbourhood as Vienna City and the Regional Police Commander himself were interviewed about sex and crime in Vienna City. Both interviewees emphatically rejected the claim that the sex industry has led to an increase in crime. The reasons given are consistent with what has already been argued, except that they mentioned increases in cases of real estate fraud – an issue discussed earlier in this chapter – unrelated to the sex industry. While they will not provide statistical information on the basis that the topic borders on security and the Police Service would need to issue statistics according to its own regulations, the qualitative information provided about the absence of sex-related crime in Vienna City is corroborated by the numeric data set out in Table 6.4 and discussed further in Chapter 6. In addition, taxi drivers with whom I discussed the issue of crime did not accept the posited connection between sex work and crime.

Next, there is the claim that the Vienna City area is a hotbed for poor morals. As Nate Plageman shows in his recent book, *Highlife Saturday Night* (2012), the connection between sex work and poor morals in Sekondi-Takoradi is not new, although often inconclusively proven. Yet, in the case of Vienna City, the link is particularly weak. In fact, it is a mixed use area, with restaurants, and catering services co-sharing the space. In turn, families visit to eat together and use the place as a social gathering site, even on Sunday (Sabbath for most Christians in Ghana) night.

The suggestion that droves of whores have taken over the city is equally questionable. The numbers are substantial and the atmosphere charged, but they are nowhere close to how they are portrayed in the press. However, these features or functions of Vienna City are hardly ever mentioned or they are assumed away as non-existent. Thus, strictly speaking, the charge that the whole Vienna City area is a site of moral degradation ought to be problematised.

Even with the user devoted to clubbing, it is not quite accurate to assume that Vienna City is a moral disaster. I compared the type of dressing of the sex workers to what I found when I visited the university and polytechnic campuses in Takoradi, Kumasi, and Accra and found that the types of dresses had more commonalities than differences. The claim that all or most sex workers are children is also exaggerated. My visual assessment reveals that a huge number of people look mature, with braided and permed hair[4] – characteristics proven to be largely synonymous with adulthood in Ghana (Essah, 2008). There is evidence that street children are involved in sex work in Takoradi (Wutoh *et al.*, 2009), of course, but that is different from media representation that the industry is dominated by child sex workers. Another reason why the thinking that Ghanaian children have been attracted en bloc is problematic is that foreigners are very active in the sex market too. While there are apparently no non-black sex workers, unlike Cameroon where Asians, especially Chinese, are involved in sex work and have been nicknamed 'Shanghai Beauties' (Ndjio, 2009), there are a large number of people of black race descent, coming mainly from Benin, Côte d'Ivoire, Liberia, Togo, and Nigeria all involved in the sector, alongside the Ghanaian sex workers.

Further, the suggestion that temporary migrants have suddenly introduced a moral ill into the metropolis is contestable. Ghanaian historian Emmanuel Akyeampong has shown that sex work is neither new nor a recent phenomenon in Sekondi-Takoradi or Ghana. Indeed, as far back as the 1930s and 1940s, sex workers were in substantial numbers in Sekondi-Takoradi and at a point were even regarded as 'proper women' for marriage. In spite of that, the women were nicknamed after commercial shopping outlets such as UAC and Leventis to signify how commercial were their services (Akyeampong, 1997). According to Busia (1950), the sex workers sometimes had 'pilot boys' who piloted clients to them. The pilot boys themselves seemed to favour certain clients especially the American sailors and soldiers because they paid them better. The key point, however, is that, even if there has been a *surge*, sex work is not new in Sekondi-Takoradi and hence Vienna City cannot be blamed solely for introducing moral decadence, if in fact there is such decadence, in the metropolis. Indeed, the contention that this part of the city undermines family values ignores a key function played by the restaurants in Vienna City: serving as a place of relaxation for families.

While the concern about HIV prevalence rate in the region in which the metropolis is located is gaining wide support (see, for example, UNDP, 2013), it lacks empirical support. The scientific evidence presented by the Ghana AIDS Commission (2012) does not support it. While there was a rise in HIV prevalence rate from 2.9 to 3.1 between 2008 and 2009, the rate has since then consistently declined: from 3.1 (2009), to 2.5 (2010), and then to 1.9 (2011). Indeed, the Western Region is not one of the HIV 'red alert' zones – Central, Eastern, Greater Accra, Ashanti, and Volta – where prevalence is on the rise. Records about Sekondi-Takoradi itself are not available, but it is known that HIV is more prevalent in Cape Coast – where there is no oil.

It may be that more HIV infected people have moved into the city to look for sex or non-sex related work like other professionals. Or, one might say that the perception of high health risk has some validity, given that research on such issues in Accra by Anarfi (1997) and others (e.g. Asamoah-Adu *et al.*, 2001) confirm a high prevalence rate. However, aside from lacking empirical support, such a deductive analysis is problematic because engaging in sex work only increases the risk of spreading sexually transmitted diseases. As shown in the research by the Ghana AIDS Commission (2012) and globally established elsewhere (e.g. Vanwesenbeeck, 2011), condom use significantly reduces the risk. So, in the case of Sekondi-Takoradi where prior research (Wutoh *et al.*, 2009) has shown that even children use contraceptives, the link between the risc of the sex industry in Vienna City and the spread of diseases is tenuous. Ultimately, the discourse that HIV/AIDS is prevalent ought to be problematised and fresh empirical studies following Anarfi's work in Accra ought to be conducted.

Finally, the perception that property values may have declined because of the activities of sex workers ought to be carefully studied because it is a globally held view. Indeed, as far back as January 1914, J.C. Nichols published an influential paper on 'housing and the real estate problem' in which he claimed that building or renting a house near the location of prostitutes tended to depress house and real estate values (see Nichols, 1914, p. 136). This view has remained prevalent in real estate and planning circles to this day (Hubbard *et al.*, 2013).

Much has been written on the drivers of property values and prices in urban Ghana (e.g. Owusu-Ansah, 2012; Awuah *et al.*, 2014), but not on the impact of sex work on property values. So, as earlier explained, I interacted with estate agents, property valuers, and land professionals respectively in private professional practice and the Lands Commission in the area to have a sense of the trends in property values.

There is no evidence of depressed values. Indeed, Vienna City is located in one of the prime areas in the city. It is in a neighbourhood that features top hotels, quality security, education, and other amenities and facilities such as hospitals and quality roads that are established (see, for example, Owusu-Ansah, 2012; Awuah *et al.*, 2014) to enhance property values in Ghana. My interviews with the property market analysts reveal substantial increases in value – a trend corroborated by Yalley and Ofori-Darko (2012) and Yalley *et al.* (2012) in their study of property prices in the twin city and its surrounding settlements. This trend is intuitive and consistent with property valuation principles, given that the area is safe and serene. Whether the values would have been higher but for the sex activities is counterfactual and hence difficult to prove.

Conclusion

The experience of Sekondi-Takoradi's urbanism under oil brings to question the sweeping and categorical claims of both the old ('oil is a blessing') and new ('oil is a curse') orthodoxies. The evidence shows that oil is not entirely a blessing and the so-called 'curses' in the urban economy are not homogeneous, but class based. The process of accumulation is attained at the expense of social dislocation, and how claims of crowding out, in practice, can be crowding in as in when hitherto non-estate agents expand rather than *shift* their economic activities to become estate agents. This chapter points to the changing livelihoods in the city, starting from the rigs, their employment in construction firms, and estate agencies, to the transport sector and nightlife. There have been clear gains, but also major losses and lack of congruence between expectations of employment and the reality of limited places.

These variegated experiences co-exist. Most of the dynamic benefits are distributed along class lines and people who are most comfortable are attaining more economic gains. Chiefs, wealthy indigenes, investors from different cities who are Ghanaians and expatriates have captured the best part of the boom in economic activities in the city. The speculation, competition, and population growth from in-migration have been responsible for increases in land values and housing prices which have hurt the majority poor in terms of their access to accommodation, but have enhanced profit for developers, for example.

Overlying this mosaic of privilege is the gender question. The boom has been in male-dominant sectors, that is, areas culturally and occupationally perceived to be for the masculine gender. Where women are seemingly taking 'advantage', as in the sex industry in the Vienna City, Takoradi's Red Light District, they have been criminalised, cajoled, and personally blamed, although it is their bodies that some will argue have been exploited and hence they need support rather than attack. The issue about how to develop systems and mechanisms for these women to enter formal, better paid, and regulated employment sectors has been ignored, while the women have been individually blamed even then for unproven charges. Taking into account these difficult outcomes, the experience of economic change during the oil era in Sekondi-Takoradi questions the existing framing of the influence of oil, whether as a blessing or a curse. Indeed, the regional angle of the dynamics, including the influx of migrants from outside Ghana will further expose the 'resource curse' doctrine as a framework overly fixated on national-centric processes, while ignoring other scales whether cascading down or scaling up. These weaknesses, tensions, and contradictions have socio-environmental angles to which the next chapter turns.

Notes

1 See its website www.gogspa.org/ for activities of the association.
2 I obtained part of this information from the *Visitors' Notebook*, interaction with attendees, inscriptions on some of the cars parked, and suggestions in the speech made by the developer and owner of the Oil Village.
3 I thank Augustine Beakana, Head of Credit, Ahantaman Rural Bank, for calling my attention to this pervasive view and for sharing his own experiences with car registration.
4 Note that 'children' (including all high school students) are barred by the Ghana Education Service from making/perming their hair.

6 Fishers and farmers in a changing twin city

Introduction

It is important to ask in what ways the oil industry is shaping the ecology and agrarian relations in and around the city given the centrality of the environment to society, economy, and politics (Goodman, 2011). In the Ghanaian case, the importance of fishing and farming (Government of Ghana, 2012) makes separate study of these sectors particularly significant.

Yet, current research does not address the social, environmental, and political economic conditions entailed in the ongoing changes in the city, especially for the fishing and farming classes who are likely to experience oil exploration in different ways. Some of the oil companies claim in their environmental impact assessments prepared to support their application for licences to operate in such a way that their 'footprints' are negligible and suggest that there is no need to study it (see, for example, Tullow Ghana Ltd, 2009, p. XLV). However, institutions such as the Environmental Protection Agency seem to suggest that these claims cannot be taken on face value and research is required to monitor these impacts (EPA, 2011c). Indeed, my research at the EPA library in Accra (Ministries) and interaction with its main and assistant librarians[1] confirm that EPA has not undertaken any such study. The institution, however, has started setting up the EPA Petroleum Department (Oil and Gas Capacity Building Project) whose mandate is to track the impact of the oil industry, although its work is at a very preliminary stage with no designated office as of the time of writing.[2]

This chapter tries to fill this gap. It investigates the processes, peculiarities, and complexities of how fishers and farmers in the Sekondi-Takoradi Metropolitan Assembly in the Western Region of Ghana have experienced the oil industry. The chapter moves beyond simplistic discourses of resource curse that have dominated the discussions of oil in Ghana and elsewhere in Africa. It argues that oil spillage and resulting pollution have been minimal to date. Yet, there have been multiple forces of expropriation at work: fishing rights are under threat, land rights are under threat, and hence the livelihoods of fishers and farmers are under threat. While there is no evidence of massive spillage, pollution seems to be a Sword of Damocles. Rather than a naturalised 'resource curse', this chapter argues that the imbalance between the power of the state, the global order, and neoliberal push for 'good governance' are key factors that have shaped and constrained the status quo. The rest of the chapter is divided into three sections. The next section turns to data issues, and then to analysis. Finally, the chapter concludes in the last section.

Sources of data

From the previous chapter we can see forces of 'progress amid poverty' at play in the city, and this notion is used as an organising frame for this chapter. It is a useful perspective because it offers a fruitful amalgamation of political ecology and heterodox political economy, two of the most effective lenses to study economy, ecology, and natural resources (see, for example, Escobar, 1998; Goodman, 2011). Even so, a strict interpretation may make it deterministic. To overcome this potential danger, this chapter pays attention to the role of institutions that impinge on access to and use of land and natural resources, while simultaneously recognising the influences and limitations of broader structural forces and global political economic influences (Loomer, 1951; Castle, 1965).

The data for the analysis were generated from seven main sources, namely surveys, legal instruments, reports of state institutions, government press releases and press statements and reports of oil companies, and reports of civil society and media groups. Table 6.1 provides details of the surveys used. All of them were conducted for academic purposes in the Western Region of Ghana where oil is currently being extracted and exported. The surveys were designed to look at the effects of oil discovery on land values as earlier described, marine life, and human livelihoods (Amoasah, 2010; Boohene and Peprah, 2011; Egyir, 2012). The approaches used for these surveys are similar and the results obtained are mainly complementary. For example, data taken on incomes in one survey can be complemented by another set looking at perceptions of the effect of oil on incomes/livelihoods. All the surveys were carried out during the 2010–2013 period, making it possible to capture the changing dynamics of the oil industry and the region.

While some of the reported data on incomes may be spurious because of a well-known problem of underreporting among respondents in Ghana (e.g. Tipple *et al.*, 1997; Grant, 2009), the problem varies according to who is collecting the data. One study (Obeng-Odoom, 2010) showed that for similar questions, Ghanaian natives are better able to win the trust of respondents than 'others'. While all the surveys were conducted by Ghanaians and so may be regarded as fairly reliable, it is best to regard the survey evidence on incomes as stylised.

The main regulations, acts, and bills used for the analysis are the Dispersant Use Policy, 2009; National Oil Contingency Plan, 2010; Draft National Environmental Policy, 2012; Offshore Petroleum (Marine Pollution Prevention and Control) Regulations, 2010; Marine Pollution Act, 2010. These are complemented with reports and press releases of state and quasi-state institutions such as the National Development Planning Commission and the Ghana Maritime Authority, and press statements and reports of oil companies working in the oil

Table 6.1 Surveys on the effects of oil

Survey	Period	Sample characteristics	Source
Expectation of how oil exploration will impact on livelihood and experience of the environmental impact assessment process.	December 2011	27 respondents made up of 4 development planning officers in 4 of the districts in the Western Region, including STMA; 3 chief fishermen; 1 Regional Chairman of Ghana National Marine Canoe Owners' Association; and officials of the Ghana Private Road Transport Union (2); fishmongers (3); coconut oil producers (2); ordinary fishermen (9); farmers (2); and 1 estate agent.	Bawole, 2013
Perception of decline in livelihoods.	June to July 2011	Structured interviews with 180 people, mainly women fish traders, fishermen, and/or farmers, the youth, and chiefs in the Cape Three Points area (Ahanta West District). In addition key informant interviews were carried out with people such as the chief and the headmaster of the community either by email, phone conversation, or face-to-face communication. The data were collected with the help of undergraduate students.	Egyir, 2012
Perception of decline in livelihoods.	November 2010	Structured interviews with 240 adult women in the Cape Three Points area (Ahanta West District). The data were collected with the help of undergraduate students.	Boohene and Peprah, 2011
Ascertain the actual effects of oil on livelihoods and consider how local communities are coping. The study also tries to understand the activities of the oil companies and government to mitigate harmful potential effects.	2010 It is not clear what the precise period is.	Interview with 25 household heads (all fishermen) chosen from 3 communities in the Cape Three Points area (11), Ntakrom (7), and Akwadae (7) all in the Ahanta West District Assembly. An unnamed number of informal interviews were held with key informants, including representatives of the district assembly and the oil companies.	Manu, 2011

Sources: the data sources are indicated in column 4 to correspond with the specific surveys.

fields in Ghana such as Tullow Ghana Ltd and KOSMOS Energy. Because the media has been an active part of the discourse on the oil industry in Ghana, I draw critically and selectively on the credible press, guided by the comprehensive analysis of press standards by Margaret Ivy Amoakohene (2006). In spite of these efforts to 'clean' the data, there are inherent problems such as selective bias.

The reports and press releases present a different set of challenges from the surveys. Most are not research studies and tend to present the particular view-points of the institutions from which they emerge. Also while state institutions and the state-owned media houses in Ghana have considerable merit, compared with others in Africa (Gyimah-Boadi and Prempeh, 2012), there is the tendency to present the government in the best light possible. Yet, there is an open space for divergent views and for the survival of civil society groups that report issues that are not always palatable to the government (Farouk and Mensah, 2012). Thus, by considering all the available evidence, through a process of data trian-gulation (Oppermann, 2000), a conceptual lens of 'progress amid poverty' as explained in Chapter 2, this chapter tries a synthesis which the individual data sources in and of themselves cannot readily provide, from a heterodox urban political economy perspective.

Offshore fields, institutions, fishers, and farmers

Currently, Ghana has four sedimentary basins (Côte d'Ivoire Tano Basin, Central (Saltpond) Basin, Accra/Keta Basin, and Inland Voltaian Basin) where there are various oil exploration, production, and development activities going on (Amoasah, 2010). It is in the Jubilee Fields within the Cape Three Points sub-basin located in the Côte d'Ivoire Tano Basin where the most economically viable work is being conducted (Amoasah, 2010), as briefly mentioned Chapter 3.

The oil industry has a large array of actors. The oil companies operating in the Jubilee Fields are as follows: Tullow Ghana Ltd; Anadarko Petroleum; KOSMOS Energy; Ghana National Petroleum Corporation (GNPC); and Sabre Oil and Gas. Institutions such as the Environmental Protection Agency (EPA), the Geological Survey Department, the Minerals and Mines Inspectorate, the Ghana Chamber of Mines, the Ministry of Mines and Energy, and the Minerals Commission are currently involved – in various ways – in the regulation of the extractive industry (see Taabazuing *et al.*, 2012 for a discussion of their respec-tive roles), but institutions such as GNPC, the Ministry of Energy, and EPA have more direct responsibility over the sector. At the urban level, it is the Sekondi-Takoradi Metropolitan Assembly (STMA) that has the responsibility for ensur-ing that oil is used in such a way that urban life will be enriched.

It is important to ask an oft-ignored question, namely what has been the socio-ecological cost entailed in the rise of the oil and gas industry. Addressing this question poses significant difficulty because the industry is fairly new with little information about it and its relationship with broader society. Yet, on this specific question, some suggest that it is possible to attempt a tentative answer by drawing inferences from the gold mining industry. Gold mining employs

15,000–18,000 people in Ghana, less than 1 per cent of the total labour force. Yet, the operations of only two companies, Newmont Ghana Ltd and AngloGold Ghana Ltd have displaced over 50,000 people who have been poorly compensated (Owusu-Koranteng, 2008). A comprehensive study of the effects of gold mining in Ghana undertaken by the Operations Evaluation Department (OED) of the World Bank acknowledged, rather solemnly:

> It is unclear what its [the gold mining industry] true net benefits are to Ghana. Large-scale mining by foreign companies has high import content and produces only modest amounts of net foreign exchange for Ghana after accounting for all its outflows. Similarly, its corporate tax payments are low, due to various fiscal incentives necessary to attract and retain foreign investors. Employment creation is also modest, given the highly capital intensive nature of modern surface mining techniques. Local communities affected by large-scale mining have seen little benefit to date in the form of improved infrastructure or service provision, because much of the rents from mining are used to finance recurrent, not capital expenditure.
>
> (OED, 2003, p. 23)

There is strong corroboratory evidence (e.g. Akabzaa and Darimani, 2001; Yelpaala, 2004; Hilson and Banchirigah, 2009) that in gold mining towns such as Obuasi, Tarkwa, and Akwatia, many people are landless, homeless, unemployed, poor, and weak. Many have died as a result of the devastating impact of mining on the environment. Such devastation sometimes leads to protest by the mining communities, often led by farmers who stand the risk of losing their farmlands. In this regard, there have been protests in Obuasi and Asunafo North lately. But, as noted by Owusu-Koranteng (2008), there is a lack of community based organisation in many mining communities, which has led to the emergence of some civil society organisations like the Wassa Association of Communities Affected by Mining to help in the organisation of struggles and protests against the mining companies. Such struggles have pressured the companies to make minor changes in their operations and have also created national awareness on the dramatic devastation of mining companies (see Owusu-Koranteng, 2008).

While this evidence is strong, it does not follow that gold mining does not support the Ghanaian economy as might be thought. Indeed, the work of Bloch and Owusu (2011, 2012) has systematically shown the vast contribution made by the industry through its forward and backward linkages to Ghanaian society and economy. The industry has contributed substantially to the development of ancillary industries such as input making and supply industries involved in electrical and plastic work. Some mining companies have strict guidelines on obtaining its suppliers from local companies. Further, mining companies in Ghana typically commit to corporate social responsibility. So, the popular perception that gold rents are trapped in an enclave such that society has not benefitted at all is contestable.

Nevertheless, it is wrong to dismiss the growing concerns of local people about the gold mining industry. Indeed, recent research by Taabazuing *et al.*

(2012), employing ethnographic methods, note that while there are some benefits arising from gold mining in Wassa, where the study was conducted, on a net basis, gold has done more harm. The study shows that gold mining has been carried out with little or no compensation, inadequate employment opportunities for locals, growing distrust between ordinary community members and mining companies and government officers.

It is tempting to conclude from this evidence that oil will have the same impacts. Although both are extractive activities, heterodox land economists typically argue that places, resources, and institutions are different and hence even similar processes will generate different political, ecological, economic, and social outcomes (e.g. Loomer, 1951). According to Basedau and Mehler (2005), it is incorrect to assume that negative ramifications of different resources play out in the same way. In the Ghanaian case, the dynamics of gold are substantially different from oil. The gold industry is a product of colonialism. It grew with the city of Obuasi, for example, whereas oil is a relatively recent find, in a city unfamiliar with oil revenues. Further, the oil resource is offshore, gold is onshore.

Also, the gold industry does not entail as many consultants, and such huge infrastructure as the oil industry. Finally, the dynamics of oil seem to be simultaneously local and global, whereas gold is largely exported and its dynamics play out quite differently from oil. It is for this reason that new institutions have been set up and developed for the oil industry, including new laws, new departments, new funds, and new consultants. Thus, the question, at what cost is oil accumulation, especially for fishers and farmers and the environment in the Western Region of Ghana, looms. Although EPA has not conducted its own research into the issue (as revealed by my research at its library and interviews with some members of staff of the Petroleum Department), it claims that 'Ghana's offshore petroleum industry has had no significant effects on the marine and coastal environment for now due to minimal activity.' In what follows, however, a synthesis of available piecemeal evidence, suggests that EPA's position is overly optimistic and inconsistent with 'conditions on the ground' (EPA, 2011c, p. 4).

Socio-ecological impacts

Sekondi-Takoradi has a vibrant fishing industry which is linked to its entire urban economy and society. The Western Region produces one-third of the fishing harvest in the country, although it is one of ten regions (Coastal Resources Center, 2010). The industry constitutes a complex food chain, made of fishers (often men), traders or mongers (often women), retailers, and producers. Also, there are those actors at the base of the chain who supply inputs to the fishing industry.

The fishing industry is crucially important to Ghanaians, mainly because fish is such a core part of Ghanaian diet, constituting some 40–60 per cent of animal protein (Gordon *et al.*, 2011). Outside Ghana, the fishing industry has attained some influence too. A small market for fish is available for the industry in neighbouring countries such as Togo and Côte d'Ivoire. Fishing and water resources in the Western Region constitute a major part of a rich and diverse biota. Indeed, the region is widely regarded as containing more biodiversity than any region in Ghana (Coastal Resources Center, 2010). There are numerous fresh water bodies, wetlands, different plant species, and rain forests in the Western Region. Indeed, the Cape Three Points Forest Reserve in the region is the last protected coastal forest in the entire Gulf of Guinea, and the Ankasa National Park, the most biologically diverse terrestrial site in Ghana, can be found in the region too. The nine major rivers, most of which connect with lagoons, flow in the region and are naturally protected with swamp forests. Huge wetlands, such as the seasonally flooded Amansuri, a well-regarded tourist site in Ghana and internationally, are all in the region (Coastal Resources Center, 2010, p. 16).

Yet, most fisher folks in the region are poor, uneducated, and dependent on oil only (Gordon *et al.*, 2011) and, since 2007 fish harvests have declined, with serious implications for the livelihoods of fishers and farmers (STMA, 2011a). The vision of the policy of environmental management of the Government of Ghana is:

> to manage the environment to sustain society at large. This vision is based on an integrated and holistic management system for the environment in Ghana. It is aimed at sustainable development. The policy seeks to unite Ghanaians in working toward a society where all residents of the country have access to sufficient and wholesome food, clean air and water, decent housing and other necessities of life. That will further enable them to live in a fulfilling spiritual, cultural and physical harmony with their natural surroundings.
>
> (Government of Ghana, 2012, section 1.1)

This vision clearly shows that the oil industry ought to be managed in such a way that everyone wins. The vision tries to adopt a conflict free orientation, but does this reflect the conditions on the ground? While it is theoretically possible that the growth of the oil industry can create opportunities for fishers and farmers, currently there does not seem to be substantial evidence to demonstrate this positive linkage. Available evidence points to issues of spillage and enclosures with implications for fishers and farmers, and hence all the people involved in the food chain described earlier. These issues require more careful analysis to which the chapter now turns.

Spillage, enclosures, and fishers

A ministerial committee was set up to investigate allegations that KOSMOS Energy, one of the oil companies operating in the Jubilee Field, had spilt 706 barrels of toxic substances into the sea, causing severe environmental damage. The committee found the company culpable and, accordingly, fined it $35 million. The company is on record to have undertaken to be a lot more careful in its operations, and, since the last spill, there is no clear evidence of a repeat of the 2010 incident (KOSMOS Energy, 2010). Indeed, the case of KOSMOS is unlikely to be widespread. Tullow Ghana Ltd, one of the huge oil companies, for example, has generally had safe operations to date, leading the company to boast that 'it has set the industry benchmark for deep water development' (Tullow Ghana Ltd, 2010c, p. 2). Indeed, the state has put forward the Offshore Petroleum (Marine Pollution Prevention and Control) Regulations, 2010, which contains a lengthy list of what is required to be in place before it grants oil companies the right to drill, including imposing stringent requirements on the use of modern oil equipment such as sludge tanks and oil filtering equipment (Regulations 16 and 17). Recent interviews (Kombat, 2013) with the captains of key oil and gas companies in Ghana confirm that the oil companies continue to tout the sophistry of their equipment and how well they meet the local and global standards imposed on them.

However, there are strong grounds to be sceptical of these environmental successes. First, continuing external scrutiny of the compliance claims made by the oil companies is not always available. Second, while the state has put in place elaborate guidelines such as the National Oil Spillage Contingency Plan (Government of Ghana, 2010c), and the National Oil Spill Response Dispersant Use Policy (Government of Ghana, 2009), they are mostly curative in orientation, not preventative. Third, the agencies in charge of prophylactic management are sometimes rendered structurally ineffective. For instance, Samuel Marful-Sau (2010), a justice of the Appeal Court of Ghana, has argued that the GNPC finds itself in a conflict of interest situation because it is simultaneously an oil broker and a regulator of the industry for effective ecological management.

There are other reasons for which 'success' in environmental management should be viewed sceptically. For instance, the institutions mandated to ensure that these environmental standards are implemented have rather weak systems in place to enforce their own standards. The EPA, in particular, is poorly resourced and lacks the key instruments to put the plan into effect (Marful-Sau, 2010; Taabazuing *et al.*, 2012). In addition, the green movement has recurrently argued that 'ecological modernism' – the ever increasing use of so-called modern equipment for environmental management – has not been proven to substantially reduce oil spillage. Rather, the problem is moved around to other areas and other forms of emissions which are equally injurious to the planet (Salleh, 2011). Not surprisingly, most people interviewed in oil communities do not believe the EPA has the needed capacity to execute its responsibilities (Egyir, 2012).

The Ghana Maritime Authority is better resourced to ensure cleaner and greener production. Fortunately, the Marine Pollution Act 2010 empowers the authority to make sure that the provisions of the act are met. It is 'AN ACT to provide for the prevention, regulation and control of marine pollution within the territorial waters of Ghana and other maritime zones under the control of Ghana and for other related matters' (Preamble, emphasis in original). So, if the Authority succeeded in doing so, it would bode well for marine life. Relative to the EPA, the competence of the authority is not in doubts. Indeed, in 2011, it successfully conducted 100 per cent of port state inspections on tankers loading oil at the Jubilee Offshore Terminal. It found that all the tankers (20 in all) brought in to lift oil had 'high safety standards' (Ghana Maritime Authority, 2011, p. 16). However, the authority has given strong signals that it does not have the financial and human capacity to monitor all areas of possible spillage and injury to marine life (Ghana Maritime Authority, 2011, pp. 25 and 42). Further, its remit is not as wide as the EPA which is supposed to be playing a leading regulatory role in the oil industry.

In the face of these institutional weaknesses, there has been considerable interest in two policy instruments. The first relates to the use of environmental taxes and the second to self-regulation. Both are market based policies, however. Regarding the first, environmental taxes are expected to work by encouraging oil companies to adopt, innovate, or switch to clean production technologies (Kombat, 2013). Theoretically, the incidence of such taxes will be more or less depending on how polluting are the activities of the companies relative to the stated thresholds (for a general discussion of environmental taxes and the theory behind them, see Stilwell, 2012b). However, due to design defects of the environmental tax regulations and the structural reason that the companies find ways around the market that are difficult to be internalised within market trans-actions, this policy instrument has not worked to date (Kombat, 2013). Indeed, under 'Environmental Taxes', the 2013 Budget Statement of Ghana (MOFEP, 2013) does not mention environmental taxes related to oil at all.

From a self-regulation perspective, too, the state has made some conspicuous attempts. For example, the EPA (2011, p. 4) notes: '[t]he industry must be com-mitted to ensuring that the marine environment in which it operates is managed sustainably for future generations'. There are important challenges with this policy stance too. First, it is difficult to accept uncritically that the oil companies will impose on themselves stringent environmental standards that would increase the cost of production. Second, according to Amoasah (2010), there are serious problems with the methodology used by the oil companies for impact assess-ments. He argues that most of the existing methods do not take into account seismic issues and questions of noise, which according to native custom disturbs the sea gods (Boohene and Peprah, 2011). Further, there has been little analysis of the size and effects of spillage and depositions that took place during the exploration phase. Together, these examples provide sufficient evidence to dismiss claims of environmental success in the management of oil resources.

More fundamentally, there are strong grounds to argue that the drilling and production of oil have led to serious environmental problems. First, a team of sci-entists from the Department of Oceanography and Fisheries, University of Ghana, and Department of Nuclear Engineering and Material Science, National Nuclear Research Institute of the Ghana Atomic Energy has discovered high levels of con-centration of chemicals in the Jubilee Oil Field that threaten to wipe out biodiver-sity in the waters of the Western Region and the lives of its fisheries reserves (Nyarko *et al.*, 2011). Second, since 2008, 16 whales have died and been washed ashore. While the debate about whether the deaths are linked with oil activities offshore is ongoing, there is evidence that the environmental impact assessment conducted by the oil companies prior to the commencement of production did anticipate such deaths. A fisheries impact assessment should have been done by now as part of the conditions the companies had to fulfil for oil production (as spelt out in the Fisheries Act 1625) but, to date, this requirement has not been met (Aklorbortu, 2013). Third, an earlier study (Boohene and Peprah, 2011) also documents widespread fears of possible deaths of coconut trees arising from chemical absorption from oil-related activities and their resulting health impacts on local communities as coconut is part of the diet of locals. These fears are not unfounded as the official expectation seems to be that the activities of the oil companies are expected to generate up to 140 metric tons of both hazardous and non-hazardous waste on a weekly basis (UNDP, 2013, p. 96).

Civil society groups have tried to keep a watchful eye on the industry. Currently there are 150 private radio and 20 TV stations all over Ghana that are all interested in the effective management of oil (Gyima-Boadi and Prempeh, 2012). However, most of these are primarily concerned with whether there is transparency in the award of contracts (Gyampo, 2011). While important, that focus ignores contentious issues such as the decision by the state to ban, since the commencement of drilling activities, fishing in areas of the sea in order to protect oil installation and subsidise pre-mix fuel of fishers.

However, such state decisions, often made in cahoots with oil magnates, can pose a serious threat to fish harvest. One survey of the opinions of 204 female fishmongers conducted before production began (Boohene and Peprah, 2011) found that 52 per cent of them anticipated a reduction in fish supply. Another conducted after production began in the same area showed that only 36.2 per cent of the interviewees perceived a serious threat, a decline from the earlier survey, and 5.1 per cent could not directly relate their fishing with oil and gas (Egyir, 2012). This evidence seems to vindicate the position of the Ghanaian state that the effect of oil on fishing is more to do with perception than reality. Indeed, the Acting Executive Director of EPA has argued that oil installations that prevent fishing must be regarded as enhancing 'sustainable' fishing, as such installations prevent overfishing. In turn, he argues, the stock of fish will increase and hence more fishing jobs will be available (Badgley, 2011).[3]

While the ban is likely to protect life, because the exclusion zone will become a 'haven' for fish, some of which are fingerlings and so need the protection, the claims of the EPA officer are contestable because the ban will simultaneously reduce fishermen's livelihoods. This position has been persistently argued by the Line Hook Canoe Fishermen Association (Badgley, 2011). A 'before and after analysis' of incomes in Cape Three Points, where the predominant occupation is fishing (Manu, 2011) shows that the oil exploration, production, and development is eroding fishermen's incomes, as shown in Table 6.2.

Table 6.2 shows that oil exploration, production, and development has dramatically decreased the income of fishermen at all income levels. The imposition of the ban has resulted in the disappearance of the cohort of fishers that had previously declared earning over Gh¢10,000, and many appear to have

Table 6.2 Net annual income of fishermen in the Cape Three Points area

Annual income (range) (Gh¢)	Share of respondents (25 household heads)	
	Before (%)	After (%)
+10,000	28	0
7,000–9,000	9	2
4,000–6,000	28	44
1,000–3,000	8	40

Source: Manu, 2011, p. 48.

joined the ranks of other lower income brackets, swelling the share of people earning between Gh¢4,000 and Gh¢6,000; Gh¢1,000 and Gh¢3,000. Amoasah (2010) predicts a worsening of the situation due to possible flaring, spillage, and noise, air emissions, hazardous solid waste, and poisonous water discharges into the sea. The affected fisher folks will have to move out of fishing and seek alternative sources of livelihood either in another industry or in the same industry but in different locales.

The oil companies recognise that this experience presents potential grounds for conflict as fishermen may be attracted to the oil installations which, by their design, tend to dam the water in which they are installed. While fishing in and around oil installations may damage equipment, the greater risk is that the fishermen are likely to be injured or lose their lives and property (Tullow Ghana Ltd, 2010b). To avoid such problems, Tullow Ghana Ltd, for example, has established what it calls a 'collaborative approach' which entails, '[e]ngagement with fishermen and Directors of Fisheries – a collaborative approach to identify risks to both parties and to seek workable, mutually acceptable solutions'. These 'acceptable solutions' include:

> [w]orking with the Ghana Navy and Maritime Authority [which will be contracted for:] patrolling and policing fishing activity at offshore operating locations in order to 'deter fishing intrusion' [while using] Long Range Acoustic Devices (LRAD) ... to detect and deter fishermen.
>
> (Tullow Ghana Ltd, 2010b)

However, this way of tackling the issue is clearly militaristic and is punishment rather than rewards based. The approach says nothing about the reduction in fisher folks' livelihood and is more about protecting the company's installations than underwriting the survival of fisher folks. As noted by the company:

> we undertook briefings about the dangers of fishing near our operations and reinforced this education drive through poster campaigns and local radio programmes. These are in addition to the daily contact our community liaison officers have with the communities near to our operations.
>
> (Tullow Ghana Ltd, 2010a, p. 65)

Bawole's (2013) survey, however, shows that the community meetings tend to be mere window-dressing, a process whereby the oil companies and their officials talk at, talk about, and talk to the communities, never talk with them. Such meetings offer the companies the opportunity to be heard by the community, although community concerns about fears that they will lose their livelihoods, are rarely heard on such occasions. While meetings are organised by the oil companies pursuant to the provisions in sections 5 (1c) and 12 (k) of the Environmental Impact Assessment Regulations, which state that community concerns should be reflected in the oil companies' operations, in practice the meetings are basically a forum for the companies to lecture the communities rather than hear their concerns. Even when the communities voice their concerns, they have no way of knowing whether the revised strategy of the companies capture their worries. Indeed, the regulations do not clearly specify how the companies should demonstrate how they achieve this end to the communities. The EPA has the mandate to ascertain the adequacy of the revised plan but, in practice, they do not seem to do so. Indeed, there is a feeling of differentiation in the dealings of the oil companies in the sense that they do not give local planning officers sufficient attention, preferring, as it were, to deal mainly with Accra based EPA officers. In turn, grassroots participation has effectively been silenced (UNDP, 2013), or, better still, isolated.

In fairness to the oil companies, apart from adjoining communities such as those in the Cape Three Points area where they are yet to directly engage local people (see Andrews, 2013), they do organise community meetings during which they usually circulate the environmental impact assessment documents. However, as Bawole's (2013) survey shows, some local government officers and local people do not bother to study the reports, partly because they are too technical and partly because there is a wide perception that their inputs do not matter or that the 'big men' have already decided in cahoots with the government on what to do. The general feeling is that oil exploration and exportation are so crucial to the country's development that it seems every procedure put in place as a check is presumed to be a mere administrative check and that the oil companies must not be impeded in what they do (Anning, 2013, p. 230). Moreover, the local engagements do not account for possible damage to fishing facilities by support and supply vessels of oil, especially those going to and from Takoradi, which the oil companies themselves acknowledge as a potential risk (Tullow Ghana Ltd, 2009). What is becoming increasingly evident with the operation of the oil companies is that they are subordinating the local interest to satisfying the bureaucracy in Accra.

There are identifiable environmental problems that are not being addressed. For instance, local realities about the growth of sargassum (free floating seaweed), and how it reduces fish harvest, and the widespread perception that oil drilling is the primary contributor to the spread of this seaweed are not effectively addressed (Ackah-Baidoo, 2013). Neither the state nor the oil companies take the consequences of essentially side-lining the fishing communities seriously. Yet, the oil industry is engendering a destabilisation in the social order and, to some observers (e.g. Ackah-Baidoo, 2013), the failure by the industry to engage seriously with community is a recipe for widespread conflict. Already, crime statistics about the Western Region (see Table 6.3) shows that the total number of reported crimes in the region seems to have increased between 2007 when oil was found and 2010 when drilling started (81,359) compared to four years (2003–2006) before the announcement of the commercial discovery of oil (65,029). The average annual crime statistics shows figures in the 20,000s (20,339.75) compared with the period before when the figure was in the 16,000s (16,257.25).

The statistical information from the Ghana Police is not very helpful, especially because it lumps up so many different crimes in broad and vague categories. A scrutiny of the breakdown of offences in the type 'A' category does not show any direct connection with fishing industry and its experiences with oil. The all-embracing category (other offences or type 'B') is similarly unhelpful. Interviews with a Police Constable and the Regional Commander of Police (Western Region), however, shed some more light: most of the crime increase is attributable to conflicts and fraud in the land and real estate sectors. This oral evidence is corroborated by documented evidence I collected from the Rent Office in Takoradi and to which I shall return in Chapter 8.

There is also empirical research (Agbefu, 2011) to show fishers clashing with officers of the Ghana Navy who seek to protect the oil rigs from wandering fishers and these fishers typically claim they have not been well briefed on the ban enforced in such areas. More research is needed on this topic, but bearing in

Table 6.3 Crime statistics in the Western Region of Ghana, 2003–2010

	Year (with oil)					
	2007	2008	2009	2010	Total	Average
Total no. of offences (type 'A')*	18,434	19,359	22,513	21,053	81,359	20,339.75
Other offences (type 'B')**	2,568	2,543	3,542	2,256	10,909	2,727.25
	Year (without oil)					
	2003	2004	2005	2006	Total	Average
Total no. of offences (type 'A')*	14,512	15,148	17,043	18,326	65,029	16,257.25
Other offences (type 'B')**	1,581	1,956	1,603	2,525	7 ,665	1,916.25

Source: calculations based on Aning (2013).

Notes
* They can be any of the following offences: murder, attempted murder, manslaughter, threatening, causing harm, assault, robbery, stealing, fraud, unlawful entry, causing damage, dishonestly deceiving, abortion, rape, and defilement. The rest are possessing dangerous drugs, possessing Indian hemp, abduction, extortion, forgery, falsification of accounts, smuggling, possessing cocaine, possessing heroin, counterfeiting, issuing false cheques, child stealing, and illegal gold mining.
** Any other offence is type 'B'.

mind the land and housing conflicts earlier discussed (and to which I will make further comments shortly), poor community engagement, loss of livelihoods, unmet expectations of employment by the youth, there are strong grounds to argue that the increased statistics on conflict in the region has silent oil undertones. Indeed, the report of Aning (2013) shows that there are narratives of youth groups – some with connections to former Niger Delta area activists and a locally minted group by the name Cape Militia which seeks justice in the oil industry for the youth, especially. That said, we know from recent research by Paul Ugor (2013) in the Niger Delta area that armed conflict is not the only option for marginalised and overlooked groups in oil economies: they can also become anarchists or, put more mildly, they can create a shadow economy for themselves where they take actions outside formal channels of engagement with oil companies and the state.

The state and the oil companies seem to have a consensus that some 'support' ought to be given to the fishers. Providing subsidy for the purchase of pre-mix oil to be used for fishing activities is the unquestionable political promise for fishers, and, since 1990, there has been a policy to subsidise the price of pre-mix. Oil revenue did not play a role in the commencement of the policy, but it does help to sustain it. But is providing subsidised pre-mix fuel a panacea? While material support of any kind will clearly ease the financial burden on fishers, this benefit has to be placed in context. The spillage of pre-mix fuel tends to cause major ecological problems and they endanger Indigenous fishing practices widely regarded to be sustainable. In turn, chief fishers have often been cautious in recommending the use of pre-mix. Further, there has been widespread abuse of the intended use. Between 2001 and 2008, people tried to profit out of the distribution of pre-mix, possibly contributing to overfishing, as the number of sales points increased substantially from 128 to 900 (Ministry of Food and Agriculture, 2011).

Farmlands, expropriation, compensation

The potential impact of oil on agriculture was discussed briefly in Chapter 3. Of course undermining fishing will have an impact on the viability of agriculture generally because of the structural connection between fishing and agriculture. Farmers are another important grouping in agriculture in the metropolis. Fishers are also farmers, of course but by 'farmers', as used in this context, I refer to those who *till* the land not the 6 per cent of the metropolitan population in agriculture who farm the sea (STMA, 2013). Currently, 21 per cent of the population in the region (that is some 84,849 people) farm plots of land which average five acres and produce at the subsistence level (STMA, 2011b, 2013). Table 6.4 summarises the data on some of the crops that are grown in the metropolis.

Overall, 34 per cent of the cultivable land in STMA is under farming (STMA, 2011b). More recent comparable figures are not available, although indicative statistical information based on 2010 figures in the *Western Region Human Development Report* (UNDP, 2013, see p. 165) suggests that there has been

Table 6.4 Major crops and their tillage (in hectares) in the STMA, 2001–2005

Crop	2001	2002	2003	2004	2005
Maize	586	608	620	1,154	1,270
Rice	96	141	140	184	193
Cassava	2,091	2,194	2,190	2,210	2,531
Yam	28	36	40	41	33
Cocoyam	118	134	140	138	110
Plantain	227	339	140	342	359

Source: STMA, 2011b.

much reduction in agrarian activities. Nevertheless, what we learn, overall, is that any impacts on farming and fishing will have far reaching implications for a wide variety of people who depend on these subsectors of agriculture.

Certainly, for the active population of the region, the 14–49 age group, agriculture is a major employer. It provides employment for 18 per cent of the population. Of this share, some 5,000 people are fishers, while some 47,000 people farm plots of land which average five acres (STMA, 2011b; UNDP, 2013). In practice, the distinction between fishers and farmers is not very helpful because a large number of fishers are also farmers. In one study of 180 respondents in Efasu, a fishing community near the oil fields, 48 per cent of the respondents worked simultaneously as farmers (Egyir, 2012), therefore much of the problems earlier discussed will also apply to them. That said, there are dynamics that are experienced by farmers mainly because they are 'land based'.

First, there has been large scale land acquisition in this new oil economy. For example, the Ministry of Energy acquired 1,131.57 acres in Punpune while Cirus Energy, a private company, acquired 600 acres of land in Egyamra, both in the Western Region (Aning, 2011). Some 78–80 per cent of land in the region is under customary tenure (UNDP, 2013), so it is traditional leaders who control most land. The problem with the land transfers is potential conflicts based on exploitation but also on clash of cultures and fear that in-migrants are going to take the jobs of indigenous people (Aning, 2011). According to Aning (2011), there is no direct evidence that the sales were not 'properly done', in the formal system, at least. However, on the basis of the evidence from my interviews and that collected by UNDP (2013, pp. 4–5), it can be argued that the trustees of the land have sold land without – as custom demands – consulting other family members or other elders. That process raises the issue of compensation, especially how fairly it is assessed, how promptly it is paid, and to whom the payment is made. While valuers and land economists trained in the art of estimating compensation abound in Ghana, there is no agreement among them on how to assess fair compensation. Some would ignore the value of crops on land, others would include it, and a section would assess value without considering the value of land. The debate seems to relate to issues of double counting and matters of land use rights and how they differ when land is cropped or bare. Yet, the existing Mineral and Mining Act 2006 (Act 703) and the 1992 Constitution which guide the assessment of compensation for land taken in the process of mineral extraction in Ghana do not resolve these disagreements (Ayitey *et al.*, 2011).

Regarding the state valuer, the Land Valuation Division of the New Lands Commission, the method it uses for estimating compensation is a highly contested procedure called 'crop enumeration' (Obeng-Odoom, 2012b). By this method, value is estimated by counting how many crops are destroyed (X), estimating their level of maturity (young/seedling, medium, and mature) (Y), and assigning prices (Z) obtained from the Ministry of Food and Agriculture to them. From this perspective, compensation is estimated at $X * Y * Z$. This approach is neither supported by valuation principles nor any concepts in land economics (for a discussion of valuation methods related to compensation, see Ayitey *et al.*, 2011). In turn, mineral-related compensation in Ghana is widely perceived to be low and unrealistic (Taabazuing *et al.*, 2012). Sometimes, the compensation is wrongly paid to chiefs – a procedure W.K. Brobbey, Ghana's foremost land economist has questioned as inappropriate (Brobbey, 1990). The timing for the payment of compensation is similarly problematic, as the state takes a long time to pay (Larbi *et al.*, 2004). We shall return to the issue of compensation in the next chapter.

In the meanwhile, the exploitation, production, and development of black gold are pushing out both farmers and fishers in other ways. They are unable to cope with the escalating prices of land and housing, coupled with evictions, as described in the previous chapter. There is further corroboratory evidence. Gadugah (2009) has reported that there have been sharp increases in rental values, ranging from 20 to 88 per cent since the discovery of oil (see also Yalley and Ofori-Darko, 2012). Also, landowners have started selling off their land to rich oil companies and their employees. Edem's (2011) survey shows that most of the 'local' migrants find it difficult to obtain accommodation because of the high rental level in the city. In turn, they tend to be moving to low-income areas of the city such as New Takoradi where, according to recent research (Owusu and Afutu-Kotey, 2010) the standard of living is very low. Some landlords have refused to renew tenancy agreement for sitting tenants, and others have ejected sitting tenants, and many more solicit offers from oil companies (Enin, 2011). It has become typical for landlords to first ascertain which oil companies potential tenants work for before starting negotiations for tenancy (Enin, 2011). Admittedly, these pressures are experienced by most low-income groups in the region, not only farmers and fishers.

Yet, for farmers and fishers, the insecurity of their livelihoods is particularly substantial. Recent deliberations of the National Development Planning Commission of Ghana (NDPC) confirm the uncertain futures of the fishing community. The Commission considered the possibility of more oil discovery and hence more losses in livelihood. Fishing on the high seas was considered as an option for fishermen, but the members dismissed it because it is expensive to finance the safe but sophisticated technology needed to fish on the high seas. Apparently the consensus of the group was that fishers should gradually consider more profitable outlets, to move away from fishing (see comments by members of Group 3, section 4.4 on NDPC, 2011). While pragmatic on an individual level, this has severe implications for food security, first in the city, next the region, then the country and neighbouring countries.

With increasing speculation on land, there is serious concern about whether it will be made readily available for farming purposes. Also, the Government of Ghana is developing a $2–4 billion industrial estate in Sekondi to attract businesses and develop a $1.2 billion alumina processing plant to the city on a public–private partnership terms in which the government is contributing $100 million (IMANI, 2012). The effect of these developments on the agricultural sector is not immediately clear, but they are likely to raise the argument that land used for farming purposes in such highly urbanised industrial estates is not being put to its 'highest and best use', leading to forced evictions as a direct confrontation to urban agriculture. There is anecdotal evidence to suggest that this is likely to be the case and intensify in the future. The press is already publishing stories about people losing their land because they are unable to pay for urban land.

This complex mix of factors and forces of oil and the livelihoods of fishers and farmers introduces questions of urban economic inequality and ecological damage. Currently, income inequality in STMA is not as high as pertains in the Greater Accra Metropolitan Area (GAMA), but it is quite substantial. The income of the richest households ($17.67), mainly living in the Beach Road area, is five times higher than the income of the poorest households who live mainly in Kojokrom and earn $3.66 per day on average. The median household income in the metropolis is $4.91. Most of the households (74 per cent) earn between $60 and $150 a month (CHF International, 2010), suggesting that incomes in this city are about the same as the urban average and slightly lower than the conditions ($67) in GAMA (Ghana Statistical Service, 2008), although, of course, the household size in GAMA is slightly bigger (3.8 persons (World Bank, 2010)). There are clear signs of crude days ahead, and this danger, among others, will be expressed in terms of a rise in inequality, a direct sibling of what George calls, 'progress amid poverty'. Further, given that as with Ghana as a whole, 'many of these [marine resources] are endemic and some are endangered' (Tullow Ghana Ltd, 2009, p. 5), ecological sustainability faces a bleak future, albeit sometimes through oblique processes.

Conclusion

This chapter has tried to analyse the effects of the 'black gold' development on socio-ecological conditions, especially as applied to fishers and farmers in the Western Region of Ghana where oil has only recently been discovered.

Massive accumulation of capital has had considerable effects on the local economy through the creation of jobs, the stimulation of the built sector, increase in economic growth rates, creation of investment opportunities, and FDI injection from investors from all over the world. These processes of accumulation have, however, taken place at huge socio-ecological costs. While there is little or no evidence of widespread and continuing pollution of the waters in Western Region, there is no evidence of clear opportunities for farming or fishing either. Some fishers and farmers may obtain some of the jobs that will accompany the coming expansion.

Those who stay in their traditional occupations may derive benefits too, if the vigorous new economic activities generate a market for farm produce. However, it is similarly plausible and indeed empirically verifiable that those new economic activities compete for farmers' land. Currently, there does not seem to be substantial gains to the fishing and farming communities, based on either a direct policy or 'trickle down'. Instead, there is a strong basis to argue that the institutional arrangements put in place to regulate the oil and gas sector are weak and, at times, conflictual. This challenge plays out as an 'institutional problem', but it has a systemic side to it. The growing accumulation places a disproportionate amount of economic and political power in the hands of the oil magnates, at the expense of state institutions. Efforts by oil companies to support state institutions in the form of providing broader social services – widely regarded as corporate social responsibility – seem commendable, but they generate further contradictions for state institutions which are supposed to retain their independence if they are to effectively regulate the corporate sector, especially.

These jarring tensions and contradictions in the cleavage between the oil sector and socio-ecological conditions in the Western Region of Ghana pose several problems for theorists of both 'resource blessing' and 'resource curse' suasion. First, these frameworks say little or nothing about power relations. Second, they assume a neat binary of either blessings or curses, post-oil extraction. Yet, as we have seen, the alternative view, framed around property rights ideas drawn from George, Harvey, Mahdavy, and Nwoke, demonstrates that not all segments of society are necessarily blessed or hurt in the same way. Further, blessings and curses can co-exist, co-relate, and co-evolve. These processes are not easy associations and connections, as existing determinisms of resource analysis tend to suggest. Rather, they are mediated by institutions, their mix, character, and orientation.

From this perspective, the dilemmas about oil in Ghana are similar to those in other countries in Africa, although not necessarily identical. There is no evidence of widespread pollution, as pertains in the Niger Delta area in Nigeria, at least for now. Further, there seems to be a strong commitment to ecological sustainability, even if this commitment is not matched by capacity. Also, the hype attending the birth of the oil industry has been accompanied by increased civil society interest operating in a political space that is encouraging of divergent views. Here is a country where the so-called 'good governance' factor seems to hold the potential to curb the tensions and contradictions that envelop resource extraction elsewhere in Africa. Yet, a paradox arises: the very claim that Ghana's experience is expected to be different because of its abiding faith in 'good governance' implies a point of similarity with other African countries. The state has the onerous responsibility of regulating the fast growing corporate oil sector, partly because of its 'good governance' (read investor-encouraging) credentials that can generate similar outcomes as those pertaining in countries such as Congo. The abiding faith in international agreements on 'protection zones' around oil installations with insufficient local people's participation creates accumulation but also dispossession, *through legal means*. Claims of good governance and its benign effects on oil also overlook a key issue in oil struggles, namely the principle of derivations (those from whom more is taken, more should be given) which, to date, is missing in the options in place for fishers and farmers, especially, but also the people of the Western Region of Ghana as a whole. There are, indeed, difficult days ahead and a need for policy considerations in addressing these issues.

Notes

1 I conducted the research on 12 December 2012, as can be verified in the *Visitors' Book*.
2 Interview with the Assistant Programmes Officer and his team on 12 December 2012 at the EPA library where we met.
3 The EPA seems to be developing a reputation of uncritically accepting the corporate speak of 'investment for jobs and development'. In one case, in Ghana, unrelated to fishing, a Regional Director of the Environmental Protection Agency would not insist that companies acquiring large tracts of land obtain the necessary permits because he did not want to obstruct development (Schoneveld *et al.*, 2011). While it is feasible that the EPA wants to make its regulations 'flexible' in order to 'help' community people by helping corporate capital, the situation raises serious political economic concerns about the EPA's ability to be an independent umpire and questions the extent to which its own views converge with community resilience and sustainability.

Part III
Towards the good city

Photo 12 A fruits shop in Windy Ridge, Takoradi.

Photo 13 Building boom in Takoradi.

Photo 14 Booming market activities in Market Circle, Takoradi.

7 Compensation and betterment

Introduction

As noted in Chapter 1 and throughout the book, research on Ghana's oil industry is often framed in terms of the resource curse doctrine. Current studies typically consider whether the country will experience the problems associated with the Dutch disease (e.g. Breisinger *et al.*, 2010; Centre for Policy Analysis, 2010), volatility (Dagher *et al.*, 2010), and corruption (e.g. Gary, 2009; King, 2009). With the issues framed in those terms, these studies have tended to consider how to manage resources in a transparent manner (e.g. World Bank, 2009), how adequate existing policies are (e.g. Gyampo, 2011), and the influence and effects of global forces (McCaskie, 2008). These studies neither look at how windfalls and wipeouts are distributed within the urban economy embedded in particular institutional make up nor broader issues of local economic development, particularly the role of local institutions. Yet, understanding the dynamics of such windfalls and wipeouts and the mediating roles of institutions is crucial for policy making and harmonious local economic development (UN-HABITAT, 2008b; Alterman, 2012).

This chapter engages the principles of eminent domain and decentralisation to see how the payment of compensation and betterment can help to ensure harmonious local economic development. Such a consideration will be consistent with insights from George, especially, but also from David Harvey, Mahdavy, and Nwoke, as they all take a *property rights* approach to analysis. It argues that, while Sekondi-Takoradi Metropolitan Assembly has tried to position itself in a way to exploit oil rents for the public good, these efforts have proven to be necessary but not sufficient to trigger and shepherd the process of harmonious local economic development.

The rest of the chapter is divided into three sections. First notions of eminent domain and decentralisation are discussed, next questions of compensation are investigated, and then issues of betterment raised.

Eminent domain and decentralisation

The doctrine of eminent domain holds that compulsory acquisition with fair compensation is necessary in the process of economic development (see the

Saltpetre case). The doctrine takes a middle-of-the-road view of the polarised debate between conservatives who hold that there should be total private property in land and progressives who hold that private property should not be created in land because land is a free gift of nature (Alterman, 2012). The doctrine posits that it is desirable for both private and public property to co-exist, in so far as the state has the right to acquire private property for public benefit subject to appropriate compensation (Saginor and McDonald, 2009).

The theoretical justification for the payment of compensation is the notion of *takings*. A taking is either a compulsory physical acquisition of land or a reduction in the market value of land (Frieden, 2000). It may be classified as direct (arising – directly – from an act or a process) or indirect (as in a 'third party' loss), total (complete reduction of market value to zero) or partial (relatively minor reduction in property value). Alternatively, it can be private (a diminution in value caused by private estate developers) or public (value reduction that results from state or public execution of projects) (Alterman, 2011). Takings deprive landowners of their property rights which are often deemed to be secure because of the protection they receive from the public. For that reason, the state normally awards compensation to landowners who suffer a taking (Bromley, 1997).

The doctrine of eminent domain implies that landowners who benefit from an increase in their land values out of public investment or activities unrelated to their own exertion are required to make a payment to the public. This 'value capture', capture of 'plus value', or the payment of 'betterment' (Alterman, 2012) has long been advocated by Henry George who identified the problem of landowners benefitting from their land whose value tends to appreciate because of public expenditure (George, 1879 [2006]; Stilwell and Jordan, 2004; Stilwell, 2006, pp. 89–91; Alterman, 2012). George (1879 [2006], pp. 198–202) was against the payment of compensation to individual landlords, arguing that their ownership is, in the first place, unjust and so no compensation is due them.

However, the issues of property rights being considered in this chapter entail compensation to landowning communities not individuals[1] (unless where specified, as in where farmers lose their land to the government); and individuals benefitting from increases in land values not related entirely to their effort. Further, the issues are not just a 'one-off' compensation and the taxation of part of the ensuing short-term betterment, but broader roles of institutions in ongoing local development and the need to have local governments to have some mechanism of local property taxation in order to ensure that the local public services (e.g. roads, hospitals, schools, water, and drainage) necessary to cope with major economic developments are adequately funded. Sub-national governments are particularly important in such considerations because advocates of decentralisation hold that local governments are closer to the people, know their people's needs, and, being accountable to them, tend to have the incentives to ensure the local people live a satisfying life (Rondinelli, 1981; Smoke, 2003; World Bank, 2003).

Within this analytical framework, the next section addresses three sets of questions. The first set relates to compensation (namely (1) whether compensation is justified/required, (2) whether existing laws protect property rights by providing for compensation, and (3) whether the state, in fact, would pay compensation). The next cluster relates to betterment (to establish (4) whether a case for the payment of betterment can be made, (5) whether betterment payment is required by existing laws, and (6) whether, in fact, betterment payment would be made), and the last set to local governance (especially (7) what are the functions and powers of the local authority, and (8) whether the local government authority is able to use its functions and powers to foster harmonious local economic development).

Compensation

The previous chapters have shown that there are three types of takings. The first relates to loss of fishing rights as a result of offshore oil drilling and production; the second relates to loss in farmlands whose value can diminish because of oil-related environmental damage; and the third to physical loss of land arising from onshore oil and gas exploration. The Petroleum Revenue Management Act 2011, Act 815, anticipates such takings (see clause 25 of the memorandum). It notes in section 24, 3) that: 'Where petroleum operations adversely affect a community, appropriate compensation shall be paid for the benefit of the community in accordance with the relevant laws.' Similarly, the Petroleum Exploration and Production Law (PNDCL 84) of 1984 offers protection of property rights. It states that 'any person having a title to or interest in such land who suffers any loss or damage as a result of the petroleum operations shall be entitled to such compensation as may be determined by law' (section 6 (2b)). Further, there are constitutional guarantees of property rights (article 18 of the Constitution of Ghana) and protection against both partial and physical takings. Article 20 of the Constitution of Ghana notes: '[n]o property of any description, or interest in or right over any property shall be compulsorily taken possession of or acquired by the State' (20 (1)) unless it is absolutely necessary (20 (1a)) and that necessity is made public and backed by and done in accordance with law (20 (1b)). Even when these conditions are met, the expropriated persons are entitled to fair and adequate compensation which have to be paid promptly (20 (2a)). And, in circumstances where there is physical displacement, the state shall settle the displaced people (20 (3)). Also, the affected persons shall have a right of access to the High Court to challenge the taking (20 (2a)).

While the laws relating to petroleum and the constitutional provisions deal with physical taking or compulsory acquisition, there are other laws that protect takings in the form of a reduction in market value. The Local Government Act, Act 462 (section 56), and the Town and Country Planning Act 1963 (sections 21 26) stress the legal protection and compensation rights of landowners whose properties suffer a reduction in market value as a result of the implementation of a planning scheme, any work related to its execution, or any actions or inactions done by any persons to make it possible for the scheme to work.

Whether the state implements these laws fully, however, is not so straightforward. A search via the *Ghana Law Finder (2009)*, the Legal Library Services,

Gud 9t containing most digitised Ghana Law Reports, shows that cases involving compensation arising from partial takings have never been contested in the courts of law. For this reason, it is difficult to make a determination on whether the state would pay compensation to communities whose property rights have been *taken* in Sekondi-Takoradi since 2007. The realm of physical taking is rather different. There is plenty of evidence to inform contemporary political economic analysis. Historically, the Ghanaian state has chronically defaulted on compensation payment for physical takings. Between 1850 and 2004, the state executed 1,336 instruments to compulsorily acquire land. It did so in all the ten regions of Ghana. The regions with the greatest share of compulsorily acquired lands were Greater Accra (34.1 per cent), followed by the Western Region (26.7 per cent) where Sekondi-Takoradi is located. The state defaulted on the payment of most of the required compensation (Larbi *et al.*, 2004, pp. 121–122).

In a few instances, the state did pay compensation. However, even then, it tended to make procedural errors in its payment mainly because of the fear of offending powerful interest groups such as tribal chiefs (Brobbey, 1990). Sometimes too, the method of assessing compensation was problematic as it generated conservative estimates (Kasanga, 2001; Larbi *et al.*, 2004).

Whether the historical evidence can predict the behaviour of the Ghanaian state in the present circumstances where oil is fuelling changes in property relations is hard to say. However, there is some evidence that the state is unlikely to depart from its past behaviour. As we have seen, the appeal by a section of the Western Regional chiefs to the Parliament of Ghana on issues of compensation, relating to setting aside 10 per cent of the oil revenue for the development of the Western Region was rejected by the Parliament of Ghana. The basis of the rejection appears to be article 257 (6, emphasis added) of the Constitution of Ghana, which states:

> Every mineral in its natural state in, under or upon any land in Ghana, rivers, streams, water courses throughout Ghana, the exclusive economic zone and any area covered by the territorial sea or continental shelf is the property of the Republic of Ghana and shall be vested in the President on behalf of, and in trust for *the people of Ghana.*

The Parliament of Ghana further argued that if it granted the request of the chiefs from Sekondi-Takoradi and other parts of the Western Region, it would be setting a bad precedent for chiefs in other mining towns to follow (Gyampo, 2011).

This singular incident foreclosed discussions about how to assess the quantum of compensation – a key question that requires further consideration. Questions of setting aside a development fund exclusively for the Western Region has been dismissed by the state (Obeng-Odoom, 2012a), but recent interviews in oil communities give the clearest evidence ever that residents desire some 8–14 per cent of 'community compensation' (Egyir, 2012). The election and subsequent Supreme Court validation of John Mahama as President, reportedly an ardent supporter of a special fund for the Western Region,[2] might change the status quo, although this may probably not happen given that the Petroleum Revenue Management Act (Act 815) puts forward a different spending arrangement. In modern valuation practice, the issue of compensation can be done from market value or developers' profit perspective, negotiation perspective, or a perspective that combines the different perspectives (Boydell and Baya, 2011). It is not clear which of these approaches informed the chiefs' demand for a 10 per cent formula, especially when we know that the rate in Nigeria is 13 per cent and some residents in Ghana prefer a maximum of 14 per cent. They sought to do so by negotiation, but without engaging the oil companies and communities which are major stakeholders. That procedure raises important political economic issues, such as whether the chiefs truly represent the people, whether they have the competence to negotiate, and whether public discourse has sufficiently focused on international private capital.

In the Ghanaian traditional system, it is believed that the Indigenous chiefs represent the people on the basis that there is considerable deference to the chieftaincy institutions (Bob-Milliar, 2009; Kleist, 2011). However, there is substantial evidence that, with respect to the land question, chiefs have recurrently abused trust reposed in them (Austin, 2005; Ubink, 2007, 2008). The Western Regional House of Chiefs – mandated by Article 274 of the current (1992) Ghanaian Constitution to deal with chieftaincy matters in the region, was described as a 'house of thieves' by one of my interviewees. UNDP's (2013, p. 5) recent report on the region highlights 'the indiscriminate sale of communal lands by traditional leaders'. Additionally, it notes 'some traditional rulers have been known to grant freehold to foreigners'. Also, it is not clear whether chiefs are sufficiently skilled to negotiate with the state and oil companies. While there is considerable evidence from countries where participatory engagement is in place (O'Faircheallaigh, 2008, 2009, 2010) that Indigenous leaders are often involved in negotiations about compensation, most are not sufficiently skilled in doing so.

Another troubling feature of the compensation issue in Sekondi-Takoradi is that the activities of the chiefs ignore the role of transnational corporations. Although documented historical and oral evidence presented by Ofosu-Mensah (2011) and Bloch and Owusu (2012) shows that international private capital has consistently provided local projects in mining towns and paid some compensation to leaders of land-owning communities and the state, there is evidence that such packages are neither 'fair' nor 'adequate' given that they tend to be one-off payments, while the effects of mining on local communities are established to be continuing (Akabzaa and Darimani, 2001; see also Boydell and Baya, 2011). These considerations show that, while the residents of Sekondi-Takoradi, especially the fisher and farming folk, require some compensation, neither the state nor private capital is fully committed to its payment.

Betterment

Not everyone suffers a taking in Sekondi-Takoradi; others such as the landowning class, enjoy substantial increases in the value of their landed properties, as we have seen. As with the rest of Ghana, land is predominantly customarily owned. Paramount chiefs manage the land customarily with some 37 divisional chiefs, about 47 sub-stools, and about 113 families. It is these traditional entities that wield land-granting rights (Farvacque-Vitkovic *et al.*, 2008), not STMA or any other planning authority. Some individuals own the usufructuary interest and others hold fixed-term leases. However, several surveys (e.g. Edem, 2011; Enin, 2011; Yalley and Ofori-Darko, 2012) show that it is private landlords (or house owners engaged in the letting of houses) and customary landowners who have benefitted the most from the substantial increases in value through sales (sometimes to business interests such as speculators), leasing, and increases in rent.

According to section 24 (1) of the Town and Country Planning Ordinance (of Ghana), 1945:

> Where the operation of any provision contained in a scheme or by the execution of any work under a scheme, any property within the area of which the scheme applies is increased in value, the Minister, if he makes a claim for the purpose within three years after the completion of the work, as the case may be, shall be entitled to recover from any person whose property is so increased in value the amount of that increase.

Factors that affect increases in land values must be determined and we must ascertain whether they arise from public investment. As we have seen in the previous chapters, there are three causes of increases in land values. First, there has been a surge in the number of immigrants who have come from various cities, towns, and rural areas in Ghana, neighbouring countries and other countries around the world, and in a situation where demand is outstripping supply, rental values have tended to rise as a result. Second, are the public investments in the city (e.g. road construction and street naming and property numbering exercise) that the state has made since the oil find.

Finally, property values have been beaten up by speculation, especially by investors who are making purchases of land in the city to take advantage of actual and potential opportunities related to oil. So, it is necessary to capture the windfalls through betterment. However analysis of the cases reported in the Ghana Law Reports, using the *Ghana Law Finder (2009)*, the Legal Library Services, Gud 9t, shows that, in practice, the law on betterment has never been tested in any major case in the Ghanaian courts. Yet, landlords continue to wallow in the affluence bequeathed by the windfall, while low-income people unable to catch up with the hyper increases in rent, housing, and land values have to look for poorer housing (Owusu and Afutu-Kotey, 2010; Edem, 2011). So, while the oil oozing through Sekondi-Takoradi has breathed new life into the twin city, income distribution in this oil city has become more unequal. It is necessary to analyse how the institutions mandated to ensure broad local economic development in the city are grappling with the tensions and contradictions accompanying Sekondi-Takoradi's nascent oil industry.

According to the Local Government Act, Act 462, section 10, STMA is responsible for the 'overall development of the district [metropolis] ... and [has to] formulate and execute plans, programmes and strategies for the effective mobilization of the resources necessary for the overall development of the district [metropolis]'. However, the ability to design plans, to identify need, and to address need, is contingent on personnel, logistics, finance, and political will, not only of STMA but also the central Government of Ghana. Since 2007, STMA has undertaken major reforms in its activities. Three are particularly noteworthy. First, it has contracted a private valuer to prepare a digitised map of land values in the assembly (Farvacque-Vitkovic *et al.*, 2008). Second, there has been a metropolis-wide street naming and property numbering exercise to enhance the process of planning. Finally, the assembly has prepared a draft Sub-Regional Spatial Plan, a draft Structure Plan for Sekondi-Takoradi, and the Western Regional Spatial Development Framework. These plans collectively set the vision, objectives, and the strategy of the city authorities to attain harmonious urban development.

While the Assembly has successfully reduced the high levels of delinquency reported by Farvacque-Vitkovic *et al.* (2008) in recent times, mainly by using improved tracking systems operated by a private specialist firm, STMA remains heavily dependent on central government revenue for its activities. Internally generated funds (IGF) constitute only a small share of its total revenue, as shown in Table 7.1.

Thus, the success of its policies is heavily circumscribed by central government policies and funds. STMA can obtain development loans to finance its activities but, according to section 88 of Act 462, STMA can borrow only up to ¢20,000,000 (about GH¢2,000 or about $1,036).[3] If it desires to borrow any extra amount, it has to consult the Minister for Local Government and Rural Development who, in turn, has to get the approval of the Minister for Finance and Economic Planning.

Table 7.1 Revenue collection in STMA, 2001–2010

Year	IGF/total (%)
2001	49.2
2002	49.6
2003	41.5
2004	24.1
2005	37.4
2006	29.6
2007	22.3
2008	22.4
2009	39.1
2010	32.4
2011	33.3

Sources: RevNet Office, STMA, Sekondi, 2012/2013, fieldwork.

To finance its plans, therefore, STMA is heavily dependent on and constrained by the central government's District Assemblies Common Fund (DACF),[4] which typically makes up over 50 per cent of its total revenue. In theory, it can reap more revenue from property rates which tend to increase as property values rise. However, in Sekondi-Takoradi, there are both institutional and systemic reasons that create a schism between the theory of property taxation and its practice. The assembly is let down by a poor property registration system. While it possesses a Deeds Registry, a majority of land transactions take place so informally in Sekondi-Takoradi that their registration is the exception rather than the norm. Also, contrasted with land title registration, deed registration does not provide sufficient protection against competing claims to land, a common problem in Sekondi-Takoradi. Further, as noted by Rebecca Sittie (2006), the Chief Registrar of Lands, most people perceive the registration fees to be too expensive relative to their incomes, while others do not understand the need for registration. In turn, the property register is a poor basis for planning a robust property taxation system.

A direct flow of income from oil funds would be a more straightforward way to enable STMA to finance its planning activities more effectively and hence ensure improved conditions in the local economy. Such a scenario may be realised in three ways. First, the amount of DACF going to local governments (and hence STMA) can be increased across the board. The second option is permitting STMA to directly tax businesses that will spring up in the oil city. The third and final option is making STMA the direct recipient of transfers from the central government's oil funds.

None of these options is currently entirely feasible or effective. The first option is general, not STMA specific. That is, it is likely to benefit all other districts that are not exposed to the peculiar dynamics of Sekondi-Takoradi as an oil city. On the second option, while under articles 174 and 254 of the Constitution of Ghana, the Parliament of Ghana can legislate for local governments such as STMA to collect additional taxes or levies outside the usual metropolitan sources of finance (such as market tolls and property rates), currently, STMA has no such options or powers. So, the second option is unlikely to happen. Similarly, the third option of receiving special funds is unlikely to materialise. The official position of the Government of Ghana on how to generate, manage, and distribute oil revenue is contained in the Petroleum Revenue Management Act 2011, Act 815. According to the Act, all petroleum rents shall be deposited in the Petroleum Holding Fund based at the Bank of Ghana (section 2).

In addition to this fund, recall, there is a Ghana Stabilisation Fund, the purpose of which is to 'cushion the impact on or sustain public expenditure capacity during periods of unanticipated petroleum revenue shortfalls' (section 9). There is a third fund called the Ghana Heritage Fund which is to 'provide an endowment to support the development for future generations when the petroleum reserves have been depleted; and receive excess petroleum revenue' (section 10). Currently, the official government position (see Kwettey, 2010) is that 70 per cent of oil rents will be expended to support the annual budget. This portion is to be called the Annual Budget Funding Amount. The remaining 30 per cent is to be regarded as 100 per cent, which will then be split into a Heritage Fund[5] (30 per cent) and Stabilisation Fund (70 per cent) – as explained in Chapter 2.

So, there is no provision for any extra support to STMA. Indeed, no member of the Civil and Local Government Staff Association is a member of the Investment Advisory Committee that is established to counsel the Minister of Finance on how to invest the revenues from oil (section 30). Members on this committee should be people who have 'proven competence in finance, investment, economics, business management or law or similar disciplines' (section 31), but not in planning. Nowhere in the Act is local government mentioned. 'Planning' is casually mentioned only in five instances. Even in such instances, planning means 'a plan prepared by the National development planning commission' (section 61), not a local government. Therefore, while the principles of decentralisation suggest that STMA, as the local government of Sekondi-Takoradi, can use the windfalls from oil exploration to improve the local economy, the analysis of the empirical evidence shows less optimistic scenarios.

Conclusion

From the perspective of public policy, this chapter shows that analysis of oil should go beyond simple binaries of 'oil curse' and 'oil blessing'. An approach that looks at lived experiences and existing regulations within the framework of eminent domain and decentralisation shows that there are few winners, and many losers. These co-exist in the twin oil city of Sekondi-Takoradi. Fisher and farming groups are likely to miss out as are the majority low-income people who struggle with housing. From a compensation perspective, the corporate oil entities and the government benefit – out of non-payment; from a betterment angle, the landowning class gains. Local government can make a difference, but current evidence suggests that the local institutions only reproduce the 'elite capture' of the windfalls in the oil city.

Small, medium, and large landowners who live farther away from the exploration drilling sites make different levels of windfalls from increases in property values. Given that this latter group is not paying any land taxes and betterment, they make huge 'unearned income'. Large landowners, including chiefs who hold land 'in trust' for their people, benefit on a net basis too. Local chiefs do not suffer personally from a taking, but they gain personally from selling long-term leases of land and receive compensation payment in the instances when it is paid. Big, private capital including investors and oil companies make the most profit – this class makes windfalls from speculation which is even sweeter at a time when the existing institutions in the city do not regulate such economic behaviour.

The local government system tasked to promote local economic development within the principles of decentralisation tries to execute its role, but with so many limitations that its effectiveness is curtailed. Not all is lost, though. Taxation provides another opportunity for the city authorities to bring about harmonious local economic development.

Notes

1 Henry George ([[1879], 2006, pp. 198–202) opposed the payment of compensation to individual landlords for whom increases in land value constitute 'unearned income'. The type of compensation under discussion in this chapter is rather different, as it relates mostly to the compensation of dispossessed communities or the taking of customary land.
2 This was reported in the press. See, for example, Nathan Gadugah's piece published by Joy FM (with audio file): http://politics.myjoyonline.com/pages/news/201011/56410. php (accessed 13 December 2012).
3 The redenominated currency is Ghana Cedi (Gh¢), not Cedi (¢) which was the currency at the time that the law was enacted. The politics and psychology of the two Cedis have recently been discussed in the *Journal of Economic Psychology* by Dzokoto and her colleagues (2010).
4 Article 252 of the Constitution of Ghana says: 'parliament shall annually make provision for the allocation of not less than five percent of the total revenues of Ghana to the District Assemblies for development' (2) 'The moneys accruing to the District Assemblies in the Common Fund shall be distributed among all the district Assemblies on the basis of a formula approved by Parliament' (3). Currently, the distribution is done according to the following formula: need (poverty defined as health, education, and water deprivation); responsiveness (how well districts collect revenues); service pressure (population density); equality (a minimum, equal amount for all districts); poverty (schools in need of major rehabilitation). In practice, what is shared is 90 per cent; with the remaining 10 per cent going to members of parliament (5 per cent); Regional Coordinating Council (2.5 per cent); Ministry of Local Government, and Rural Development (2.0 per cent to use for district assembly purposes in a way that it deems fit); and DACF Administrator (0.5 per cent for monitoring and evaluation) (Quainoo-Arthur, 2009, pp. 56–57).
5 Fund set aside for use by 'future' generations.

8   Taxation

Introduction

This chapter investigates whether the authorities in the city are using their taxation powers to take advantage of increasing land and real estate development, while examining the reasons for the status quo, and resulting implications for sustainable urban development.

The taxation of urban land is one of the least used public policy tools for sustainable urban development in Africa. This is surprising given that it has long been established – admittedly mainly outside Africa – that taxation of urban land is one of the effective strategies to develop harmonious cities (UN-HABITAT, 2008b). Most urban scholars, but particularly followers of the great political economist, Henry George, argue that the cost of providing urban infrastructure can be met by taxing resulting increases in property values (George, 1879 [2006]; Stilwell and Jordan, 2004; Stilwell, 2011). Further, they contend that increases in private land values attributed to public investments and other factors unrelated to private landowners' exertion ought to be taxed away because such increases are unearned and the resulting revenue used for social investment (Ingram and Hong, 2012).

To Georgists, land taxation is easy to administer and difficult for people to escape it, as explained in Chapter 2. Thus, proponents take the view that equitable and socially efficient urban development requires the establishment of a property taxation regime. This view has become even more highly influential in recent times in places outside the West where Georgism had the most influence (see, for example, Cui, 2011, for the case of China), with Israeli planning scholar Rachael Alterman, one of the world's foremost commentators on this theme, suggesting that cities in Africa have missed out on a key resource for harmonious urban development mainly because of the unavailability of regulations dealing with windfalls and wipeouts (see, for example, Alterman, 2011, 2012).

Ghana is one of the few countries in Africa where the regulatory framework on landed property taxation is relatively well developed. According to A.E.W. Park, an expert on African fiscal laws, '[t]here can be few countries in the world which impose a wider range of taxes than Ghana' (1965, p. 162). Ghana was the first country in the Global South to develop a decentralised taxation system and its model has been described as 'successful' (Ohemeng and Owusu, 2013, p. 3): 'an example of an opportunity for innovations in normal policy making by Africans' (p. 17). So, its cities provide a useful case study to examine the proposition that the availability of a property taxation regime would lead to the creation of harmonious cities in Africa (UN-HABITAT, 2008b). The chapter is particularly timely because according to the analyses of planning theorists and practitioners such as Alterman (2012) and Fainstein (2012), Ghana's oil can be harnessed for local economic development through the establishment of a property taxation regime.

Yet, policy and media attention has, to date, focused mainly on macro-economic aspects of oil and resulting concerns about how to manage currency volatility and issues of national level corruption in the administration of oil in Ghana. Research on taxation generally abounds (e.g. Osei and Quartey, 2005; Prichard and Bentum, 2009; Ohemeng and Owusu, 2013), but the studies on the taxation of oil have been relatively few. An early study by Nana Adwoa Hackman (2009) looked at Ghana's royalty tax system and compared it with other mechanisms available to the state in taking its share of the benefits of oil production. Amoako-Tuffour and Owusu-Ayim (2010) examined the existing fiscal laws in the petroleum sector and tried to compare their effectiveness with similar fiscal regimes in Africa. Like Amoako-Tuffour, Pamford (2010) analysed the laws about taxation of petroleum. However, he restricts his analysis to the model petroleum agreement. A more recent study (Mohammed, 2013) also takes a legal route like the earlier studies but compares Ghana's fiscal regime with that of Norway – using the criteria of neutrality, revenue raising potential, progressivity, and risk, while Sasraku (2013) looks at Ghana's petroleum tax regime and its ramifications for the long-term development of its petroleum resources.

Much like the parent literature on taxation and oil (for a review, see Daniel *et al.*, 2010), however, the taxation of oil-induced rent in the urban land and real estate market has not been investigated. The questions and claims made by Georgist political economists about how urban land accumulates value and why capital gains and rental values ought to be taxed have not, as yet, received any attention. Keith Myers (2010) made some attempt to do so in the Ghanaian and Ugandan contexts, but restricted the analysis to the taxation of capital gains on pre-oil production equipment, not urban land and real estate. And, while Yalley *et al.* (2012) have tried to look at the effect of oil production on land and real estate market, their study does not examine fiscal rules and implications of how such rules are actually administered in practice. Of course it is possible to extend the argument about the taxation of land to natural resources generally and not restrict it just to urban land.

The discussion is about the concept of oil rent and the different types of rent that can be captured by oil states in the event that they operate the oil fields with private multinational oil companies – drawing on insights from the work of George, Harvey, Mahdavy, and Nwoke in Chapter 2. Without being repetitive, oil rent – also regarded as profit oil or what the Australians call profit from the sale of oil[1] – is a key element in discussing the socialisation of oil revenues. Henry George (1876 [2006]) regarded rent as the monopolisation of natural resources and, in keeping with that tradition, Georgists conceptualise rent as any income that is not strictly speaking earned. Oil rent is an extra payment which, even if legal, raises moral issues because it is a windfall – it is what is left after statutory royalty and agreed cost recovery items have been deducted (Amoako-Tufour and Owusu-Ayim, 2010).

Curiously, the interest in oil rent in Africa tends to be in how it is transparently managed, not how it is generated and distributed. As observed by Kwaku Appiah-Adu and Francis Mensah Sasraku (2013, p. 34), '[i]t appears that the development of many revenue regimes starts from the premise of how to allocate what is finally received and ignores how the monies come in'. This focus is important because opaque management and lack of accountability have contributed to widespread problems which have been experienced differently by various classes (African Development Bank and African Union, 2009). However, it does not go deep enough. The generation, distribution, and management of oil rents are all important and a sine qua non for determining questions of taxation. So, this chapter examines the nature of the regime for the capture of rents, their distribution, and their management.

It argues that, while the existence of the regulatory regime on property taxation would suggest that it can, indeed, be used to generate more revenue and bring about more equitable urban development, the *exceptions* to the framework, the calibre of the *institutions* mandated to oversee the implementation of the law, and the *social context* within which taxation operates are key impediments to success. Thus, the efficacy of taxation is obvious, but contingent rather than assured. Some of the setbacks are design defect but there are also major structural impediments to progress.

The rest of the chapter is divided into two parts. The first examines the types of revenue and rents that arise from the oil industry and then looks into how these are distributed. The second part of the chapter respectively considers opportunities, and prospects of the property taxation regime in Sekondi-Takoradi before turning to challenges.

Generation and distribution of revenue and rents

The basis of Ghana's attempts to obtain the most from the oil industry is transparency in the management of oil revenues. The Government of Ghana has placed primary emphasis on its Oil Revenue Management Act, 851 which provides regulations on how the rents from its oil reserves will be generated. The 'petroleum revenues' defined in the Act are:

> (a) royalty in cash or in equivalent barrels of oil or equivalent units of gas, payable by a licensed producer, including the national oil company or a company under a Production Sharing Agreement or other agreement; (b) corporate income taxes payable by licensed upstream and midstream operators; (c) participating interest; (d) additional oil entitlements; (e) dividends from the national oil company for Government's equity

interest; (f) the investment income derived from accumulated petroleum funds; (g) surface rentals paid by licensed producers; or (h) any other revenue determined by the Minister to be petroleum revenue; derived from upstream and midstream petroleum operations, 'public funds' has the meaning ascribed to it under Article 175 of the Constitution.

(section 61, pp. 31–32)

Some of these funds go into the Annual Budget Funding Amount and the rest to the Heritage Fund and Stabilisation Fund.

These revenue categories have been recently evaluated by Amoako-Tuffour and Owusu-Ayim (2010) who find the fiscal regime – among others made up of the Petroleum Income Tax Law, 1987 (PNDCL 188) – that is supposed to govern the generation and distribution of oil rents fundamentally problematic. They show that it places no restrictions on how much interest the oil companies can deduct from their total revenues in arriving at taxable income. So, the companies have ample room to declare insubstantial amounts for taxation purposes. This 'thin capitalisation', as it is called in the industry, looks good when compared to the second weakness: that expatriate oil company workers are exempted from paying taxes to the Ghanaian state. In addition, the regime allows transfer pricing which, in turn, enables the oil companies to shift profits to low taxation jurisdictions. Finally, the state has no strong and standardised financial clauses to bring in extra revenue to the country and in turn creates room for individual variations.

Currently, the agreements with the petro-companies entail royalty payments, as we have seen, but only at a pale 5 per cent. Even the income tax required of the companies is not consistently paid, as we shall see later in the chapter. There is no explicit legislation for the oil companies to pay profit tax, although the 2013 Government of Ghana Budget Statement expresses some interest in introducing a 'Windfall Profit Tax of 10 per cent on *mining* companies' (MoFEP, 2013, p. 75, emphasis added). The proposed tax will be on mining, not oil companies.

It is arguable that the additional oil entitlement clause helps to capture some of the fruits of good times and so Amoako-Tuffour and Owusu-Ayim (2010) rate Ghana's fiscal regime well on a progressivity criterion. Yet, there are bigger problems. The oil contracts protect the oil companies against any additional taxes. According to one such agreement,

> No tax, duty, fee or other impost shall be imposed by the State or any political subdivision on Contractor, its Subcontractors or its Affiliates in respect of activities related to Petroleum Operations and to the sale and export of Petroleum other than as provided in this article.

(Article 12, p. 50)

The

> taxes in article 12 are royalty, income tax, additional oil entitlement, payments for rental of government property, public lands or for the provision of

specific services requested by Contractors from public enterprises; provided, however, that the rates the Contractor is charged for such rentals or services shall not exceed the rates charged to other members of the public who receive similar services or rentals.

(Petroleum Agreement for Deepwater Tano, March 2006, Article 12, p. 50)

The oil companies have successfully contracted themselves out of the taxation of their rents which, on social justice grounds, should be more equitably shared with the 'landlord', in this case, the state. Compared to a number of oil producing countries in Africa, including both old and new producers, Ghana's laws are not the worst but they nevertheless do not perform favourably. As an example, the 5 per cent royalty paid to the Ghanaian state is the lowest in the world, lower than what pertains in other Sub-Saharan African countries and below the world average of 7 per cent (Amoako-Tuffour and Owusu-Ayim, 2010).

In terms of dealing with the disposal of oil assets, again, the Ghanaian state is quick to pamper oil capital, making it a 'neoliberal state' going by David Harvey's (2006b) conception. The industry has three ways to go about taxing the gains from capital. It can ignore the taxation of gains altogether, require the payment of capital gains tax by the seller and make a corresponding deduction for the buyer to arrive at a neutral situation, or only tax gains on the seller. Ghana uses none of these approaches. Instead, the state does not tax the seller on gains made and turns around to give the buyer a reduction for the full cost of the item purchased (Myers, 2010). In one case, the country lost some $30 million from this legal faux pax (Kopiński *et al.*, 2013). The overall evaluation of Ghana's oil fiscal regime conducted by Amoako-Tuffour and Owusu-Ayim (2010, p. 7) is worth quoting:

> Ghana's fiscal regime based on 'work-program bidding', has minimum front-loading charges, guarantees minimum State take, rates favourably on flexibility and neutrality, and is progressive in its basic structure. On the surface, when compared with a peer group of countries in Sub-Sahara Africa, Ghana's regime appears reasonably competitive. But the risk of revenue delay is high and the degree of progressivity is weakened somewhat by the absence of cost recovery limits, the weak thin capitalization provisions, and the weak capacity for verification and monitoring of contractors' costs and investments. Several elements of the regime are also open to contractual variation, leaving Ghana's take of resource rents subject to potential ad hoc negotiation. The complexities of the industry notwithstanding, it is useful to standardize the key features of the regime in legislation in a way that defines the scope of discretion in the contracting process. There is scope to improve government take if the expected legislative revisions guard against open ended exemptions, allowances, withholding taxes and cost recovery measures that further compromise the progressivity of the fiscal regime.

These concerns are, however, pitched at the macro level. We need to turn to the urban or micro level to study matters that arise therein. To date, there has not been any study of the urban fiscal aspects of oil, relating to the generation of revenue through land taxation, although, as we have seen, political economists with interest in rent (George, Harvey, Mahdavy, and Nwoke) and leading planning scholars suggest that this aspect of the oil industry can make or break sustainable urban development in Sekondi-Takoradi. So, the present chapter tries to fill this gap by concentrating on the planning and taxation dynamics in Sekondi-Takoradi. It explores the opportunities, prospects, and problems of the current regulatory regime, how it is lived in practice, and its implications for urban economic development.

The chapter argues that, while the existence of the regulatory regime on property taxation would suggest that it can, indeed, be used to generate more revenue and bring about more equitable urban development, the *exceptions* to the framework, the calibre of the *institutions* mandated to oversee the implementation of the law, and the *social context* within which property taxation operates are key impediments to success. Thus, the efficacy of property taxation is obvious, but contingent rather than assured. To emphasise this point, I next consider opportunities and prospects of the property taxation regime in Sekondi-Takoradi and then highlight existing problems.

Opportunities

As we have seen, there has been considerable increase in property market activities in the Sekondi-Takoradi Metropolis since the announcement of oil find in 2007 and commencement of production in 2010. According to the physical planners interviewed, most of the construction marking the boom is not officially reported to them, although by virtue of their position as residents in the metropolis themselves they are aware that there have been massive refurbishments, renovations, new builds, and change of property user (e.g. from residential to commercial). Similar comments were made by the land officers who contend that most transactions in land are not reported to the land agencies. Nevertheless, the few reported cases captured in the official data obtained from the planning and land agencies showed significant property activities from 2007 to 2012.

Although there was a reclassification of the metropolis in 2008, resulting in a reduction in the administrative size of the metropolis, the number of planning permits approved in the 2007–2012 oil era (3,161) was about 60 per cent greater than the number of permits approved in the 2001–2006 pre-oil era (1,981) – as can be seen in Table 8.1. Furthermore, the highest annual number of planning approvals ever given in the metropolis in the last decade has been in the oil years of 2007 and 2012 – again in spite of the reduction in the size of the metropolis.

Interviews at the Lands Commission, the Land Valuation Division, and the Deeds Registry Unit confirmed that there has been a boom in land transactions, most of which are attributable to oil production. Table 8.2 contains statistical information on land documents received by the Land Valuation Division for

Table 8.1 Number of planning permits approved in STMA, 2001–2012

Year	Number
2001	257
2002	290
2003	262
2004	461
2005	320
2006	391
2007	227
2008	454
2009	767
2010	495
2011	780
2012	438

Source: author's fieldwork (data compiled at STMA Physical Planning Department, 17 December 2012).

Table 8.2 Total number of documents received for stamping in Western Region, 2005–2011

Month	2005	2006	2007	2008	2009	2010	2011
January	165	147	229	284	N/A	209	160
February	91	156	308	295	N/A	219	177
March	169	199	205	249	N/A	165	N/A
April	139	176	194	163	N/A	173	270
May	163	283	377	267	N/A	151	197
June	95	218	284	230	N/A	189	250
July	138	216	N/A	294	N/A	206	291
August	122	216	N/A	238	N/A	187	302
September	180	N/A	N/A	N/A	N/A	150	221
October	157	N/A	N/A	274	N/A	199	178
November	188	N/A	N/A	231	N/A	210	244
December	146	151	N/A	195	226	122	304
Total	1,753	1,762	1,597	2,720	226	2,180	2,594

Source: author's fieldwork (data assembled from various files at the Land Valuation Division of the Lands Commission in Takoradi, 19 December 2012).

stamping.[2] It would seem that we do not need the monthly breakdowns in the table since we are dealing with yearly changes since 2007 when oil was found. However, because of missing monthly entries in the database, yearly comparisons for this particular information can be misleading. Therefore, it is useful to pay attention to all the data available and try to make sense of processes of change and continuity.

While the Land Valuation Division does not disaggregate its data to the metropolitan level, the officer in charge of the collection and collation of the data explains that a large amount of the transactions take place in STMA. From Table 8.2, it appears that the purchase of land has increased in the 2007–2011 oil period; during which average monthly documents received for processing by the Land Valuation Division were around 233 as against the non-oil 2005–2006 era during which average monthly documents received were around 171. Further, he explained that, while population surge may be part of the reasons for this increase in land transactions, that factor does not sufficiently explain the increase. Oil is a better explanation variable. To him, these physical developments and land transactions have arisen because of oil investments (those investments made by the oil companies and the government directly related to the oil and gas industry), investment for oil (those investments being made to support the oil industry and its related activities but which can also benefit other sectors apart from oil), and investment from oil (being those investments that are taking place using oil rents or petroleum revenues accruing to the state as the 'owner' of oil).

Most of the landed transactions relate to residential uses, but these 'oil houses' are distinct in form and price from earlier housing types. Generally, they tend to be of superior quality, as we saw in Chapter 5. In turn, a higher proportion of planning proposals submitted to the planning authority have been approved. Compared with 2001, when only 24 per cent of applications were approved, interviews at STMA Physical Planning Department show that, since 2007, the share of approved applications has increased tremendously, reaching 94 per cent in 2009, 97 per cent in 2010, 98.9 per cent in 2011, and 99.3 per cent in 2012.

While, as we have seen, there has been important private investment in the metropolis, three reasons – speculation, public investment in the metropolis, and population growth – drive the boom in the property market. Therefore, from a Georgist perspective, some of the increases lead to 'unearned income' which provides justification for the city authorities to capture some of these increases in value for investment in social services (George, [1879] 2006; Stilwell, 2011; Alterman, 2012; Fainstein, 2012). Whether they actually do so or in what ways they do so requires careful analysis to which we now turn.

Taxation prospects and reality

Ghana's capital gains tax and rent tax[3] regimes provide prospects for increased revenue and ability to capture the unearned income accruing to land and house owners. The first time a capital gains law was passed in Ghana was under the leadership of Kwame Nkrumah, an acclaimed socialist and first President of Ghana. Originally, it was called the Capital Gains Act, Act 289, but it was subsequently amended to Act 304. The philosophical basis of the law was to attain socialist ends: equitable distribution of public goods and the means for their production. As once noted by Nkrumah (1962, p. 103), '[i]n Ghana we believe that it is only by socialist planning that we can industrialize and transform our country'. This motive was fundamentally different from what the coloniser meant taxation to achieve in the Gold Coast when urban taxation was first introduced in Accra, Kumasi, and Sekondi, namely to assert control over land, generate revenue for the colonial administration, and generally to entrench

British imperial rulership (Fortescue, 1990). The post-colonial, Nkrumahist taxation regime was rather more egalitarian and hence did not attract the colonial epoch protest that the proposal and implementation of urban taxation generated.

However, this social basis of planning law started to change when the socialist regime of Nkrumah was toppled. Indeed, two years after Nkrumah's regime was overthrown in a coup sometimes said to be inspired by capitalist interests both locally and internationally, military forces tried to re-write relatively progressive urban development laws. While each military junta was different, generally, the period from 1966 to 1979 saw various attempts to compromise with, not confront, the dictates of capitalism in such a way that local and international capital could accumulate without hindrance. The military regime of Colonel Acheampong, widely regarded as defiant to Western forces because of its insistence on a quasi-mercantilist food production regime, for example, talked of 'the invaluable contribution of foreign capital to the survival of Ghana's economy and called for greater foreign investment' (Hutchful, 1979, p. 40). In turn, taxation laws were amended to make Ghana an 'attractive' location for foreign capital and capital accumulation. The Capital Gains Law was repealed and replaced by the National Redemption Council Decree 197 in 1967, but another decree, 347, was passed in 1975 to amend the regulatory framework put in place in 1967. Subsequent coups by forces largely friendly to the idea of leaving windfalls untaxed led to further amendments, namely the Supreme Military Council Decree 46, made in 1976, Provisional National Defence Council law 232 of 1990, PNDCL 267 of 1991, and eventually Act 592 in 2000 (Lands Commission, *c*.2000). Like other laws formulated after Nkrumah, Act 592 contained major incentives to encourage private sector participation. The report of the Expert Committee on Capital Gains Tax reported to the Parliament of Ghana that the regulation needed to be carved in such a way 'to increase investments and encourage economic expansion for sustainable growth' (Apraku, 2000, p. 247).

To be sure, there were attempts of trying to curtail excessive profiteering. In particular, a rent tax was implemented in the country around 1974 for the first time. However, similar to the capital gains tax, it underwent several amendments to make it less progressive. At its onset, it was mainly a progressive tax, with the incidence rising as rental income increased. With time, the 'radical' character was watered down. For instance, the tax free threshold increased from ¢400 (1974), to ¢2,000 (1984), and then to around ¢8,000 (1986) (Tipple, 1988) – allowing higher rent income people to escape taxation. From the 1980s to date, neoliberalism in various shades has, in essence, been intensified with the main policy focus being a key interest in attracting foreign and local capital into housing investment (Obeng-Odoom, 2012c). Indeed, the progressive taxation formula of the rent tax regime was abandoned for a flat rate in 2006. Also, the rate of tax was reduced from 10 per cent (imposed in 2002) to 8 per cent (the current rate) in the same year (Opoku, 2012; MoFEP, 2013).

Both the current capital gains law and the rent tax law are contained in the Internal Revenue Act 2000, Act 592. The capital gains provisions are spelt out in

Chapter II. The chapter is divided into five parts, relating respectively to the rate and nature of the tax, the process of becoming liable, who or what activities are liable, the calculation of the tax – entailing what goes into the formula and the exceptions, and the procedure for payment (Chapter II, parts 1–5). That too has been amended about six times. Capital gains tax is paid, among others, when people sell, exchange, or distribute chargeable items such as land and buildings or they change the use of items that were not chargeable to chargeable items. The *capital gain* on which the tax is paid is the difference between *all* of the amount obtained when the chargeable item is realised, that is, sold, or distributed; and the cost of the chargeable item, including any amount of money spent developing that item and cost expended in a process of acquiring it (sections 95–100). The one who makes a capital gain is enjoined by the Act, within 30 days, to furnish the Commissioner details of the transaction, including the location and description of the chargeable asset, parties involved, especially the full name and address of the new owner, dates, cost base, and the amount realised (section 102). The Commissioner is required, within 24 months of receiving the written information from the one making a capital gain, to make an assessment of the quantum of the capital gained and the tax to be paid on that amount (section 103). Currently, the rate of tax is 10 per cent of the capital gained (section 95).

The rent tax regulation, on the other hand, is less elaborate. Indeed, it is given only a few lines of mention in the Act. Rent tax is pegged at 8 per cent on rental value accruing from landed property (section 9). It is a flat rate, paid regardless of the rental amount. Under the Act, the rent tax is deemed 'final', meaning the one who pays it gets to deduct rent income from other earnings for the purpose of income tax assessment. The Act allows the person earning rental income to deduct the cost of servicing a mortgage used to acquire the rental premise and a sum equivalent to 30 per cent of the gross rent for the purpose of standard maintenance, and any other outgoings (section 17).

In spite of the prospects, self-defeating provisions in the fiscal regulatory regime, whimsical procedure for tax assessment, and institutional bottlenecks militate against the use of the taxation powers of the state to enhance urban development. Each of these problems is elaborated below. The capital gains tax regime exempts from taxation, gains made when the proceeds of land and buildings realised are invested in the property market within one year. It follows that speculators in the property market are not punished for the windfalls they make mainly through wilful or involuntary speculation in so far as they reinvest within one year. Also, it exempts capital gains resulting from a transfer of ownership of the asset by a person to that 'person's spouse, child, parent, brother, sister, aunt, uncle, nephew or niece' (section 101, ss. 1c). It follows that, where these relatives simultaneously find themselves in business relationships, they take advantage of the exemption. While a transfer between spouses was historically the position taken by the drafters of the law (see Obimpeh, 2000, p. 626), the parliamentary debates on the Internal Revenue Bill, as it then was, show that it was the members of the so-called social democratic government that justified the inclusion of blood relations, claiming that it was alien to Ghanaian culture to regard only spouses as 'family'. In the words of C.O. Nyanor (Member of Parliament):

> [w]e are making laws for Ghanaians not for English people. Seriously, with our extended family system here in Ghana and the close link between extended family members, apart from the spouses I think that the others mentioned in the amendment should be there, and I will support the amendment.
>
> (Nyanor, 2000, p. 629)

The problem with this view was eloquently captured by Hackman Owusu-Agyemang, curiously Member of Parliament (MP) on the side of the party that espouses property-owning democracy:

> I can understand, but I think that when it comes to, if you like, the philosophy of the whole thing, it gives a little bit of a problem.... Why should it be limited to only, as I said, blood relations? And what makes it special? Is it because the person is your aunt then she is exempt? What about a very good friend who might have done a lot more favours or contributed a lot more to whatever you have got, rather than your blood relations?... I do not believe that the mere fact that they are blood relations entitles them to any special relations than those who might have been very good to the advancement of the individual, or for the acquisition of that wealth or gain. So I have a problem with the basic underlining principle of the whole exemption.
>
> (Owusu-Agyemang, 2000, pp. 627–628)

There are major obstacles militating against the use of the capital gains tax to raise revenue and curb wanton windfalls. Thus, while MP Owusu-Agyemang withdrew his objection when he was assured that only the listed relatives would benefit from the exemption, the status quo remains hugely problematic. To give one example, a business entity can be changed within a large family for substantial capital gains, but without taxation. Further, the procedure related to reporting capital gains is whimsical. Most properties in the metropolis are built without a paper trail, as we have seen. Building permits are typically not sought for major developments, as recurrently mentioned by my interviewees at the planning department at STMA. Additionally, research (Atuguba, 2006) on the 'culture of taxation' shows that the majority of people do not find it necessary to pay taxes, among others, because they do not feel that taxes they pay are put to good use, and that the tax authorities are not personable. More recently, a nationally representative survey carried out by CDD-Ghana between 9 May and 1 June 2012 shows that 74 per cent of the population find it 'very difficult or difficult' to know the uses to which taxes in Ghana are put. Further, 68 per cent of the people surveyed find it 'difficult or very difficult' to know which taxes they should pay to the state (Armah-Attoh and Awal, 2012). Therefore, it seems that the tax regime does not operate on a quid pro quo principle. So, the requirement in the fiscal regime that persons making a capital gain from realisation should voluntarily make a written report to the Commissioner can only be whimsical. Little wonder that, in practice, only few people make such declarations.

Similar problems plague the implementation of rent tax law. Landlords are required to make declarations about rent obtained to the Ghana Revenue Authority (GRA). Yet, a number of dealings in the land market go unreported. One survey of 20 landlords in Takoradi (Opoku, 2012) showed that most people do not know about the rent tax, let alone pay it. The survey also shows that generally most of them do not voluntarily pay taxes. While the GRA (Rent Division) makes estimates on the basis of a band of comparable rental values, this procedure is problematic as it is based on expected, not actual rent. More fundamentally, the members of staff in charge of rental revenue assessment are not trained professionally to make such assessments. In Ghana, it is land economists trained in the art of valuation and affiliated to the Ghana Institution of Surveyors who profess and are known to make estimates of value through various approaches (Opoku, 2012). So, while the GRA's capacity is generally widely regarded in Africa – obviously for being the first such revenue authority established in Africa, south of the Sahara – and it has been recently expanded by the addition of the Petroleum Unit (Kopinski *et al.*, 2013), it is not clear how rigorously it arrives at its estimates of rental value.

In spite of these problems, there is cause to believe that revenue from rent taxation can potentially increase, as shown by the number of properties that have become liable for rent tax assessment (Table 8.3).

Table 8.3 shows that the number of properties that have been brought into the rental market has been increasing, albeit moderately. The average number of properties liable for tax per month was about 582 in 2010, increasing to about 623 in 2011 and then further rising to 676 in 2012. As with Table 8.2, yearly

Table 8.3 Number of properties liable for rent tax assessment in Takoradi, 2009–2012

Month	2009	2010	2011	2012
January	539	541	602	N/A
February	N/A	544	605	650
March	N/A	553	607	654
April	N/A	575	609	661
May	N/A	589	616	668
June	N/A	596	620	673
July	N/A	597	626	676
August	N/A	597	629	683
September	N/A	597	631	690
October	N/A	597	641	692
November	N/A	598	642	693
December	N/A	599	642	695

Source: author's fieldwork (data compiled from GRA, Rent Office files, 20 December 2012).

totals might suffice, but the gaps in the monthly entries would make total yearly figures without a sense of the monthly inputs misleading.

But, even with the choice of yearly totals, care ought to be taken in interpreting Table 8.3 for a number of reasons. First, the records at GRA do not show data for the years before 2007 when oil production was yet to start. Second, Table 8.3 covers the situation in Takoradi only. Third, according to the GRA staff members with whom I discussed the data, many buildings in the metropolis are constructed without permit and hence are not captured in their records. For all these reasons, a large number of increases go untaxed.

Even if the returns were higher, it would be unlikely that they would ooze into the coffers of the city authorities, as the laws make the commissioner, not the planning authority, the recipient of such taxes. While there are opportunities and prospects for Sekondi-Takoradi to be managed harmoniously through the use of land taxation, the problems entailed in the design of the laws and their implementation raise systemic and institutional challenges. Thus, using capital gains and rent taxation for alternative urban development is obvious, but not assured.

There is no extra taxation apart from income tax which the companies are allowed to reduce by charging their cost of operation, especially capital cost, against their income before tax. This is consistent with the Petroleum Income Tax Law, 1987 (PNDCL 188) and incorporated in the agreement between the state and the Jubilee Partners:

> It is the intent of the Parties that payments by Contractor of tax levied by the Petroleum Income Tax Law qualify as creditable against the income tax liability of each company comprising Contractor in its jurisdiction. Should the fiscal authority involved determine that the Petroleum Income Tax Law does not impose a creditable tax, the Parties agree to negotiate in good faith with a view to establishing a creditable tax on the precondition that no adverse effect should occur to the economic rights of GNPC or the state.
>
> (Petroleum Agreement for Deepwater Tano, March 2006, p. 52)

From this perspective, the oil companies have filed tax returns which have, to date, exempted them from corporate taxation, as the Public Interest and Accountability Committee (PIAC) determined in its last statutory assessment (see PIAC, 2013). This experience does not arise from tax evasion. Rather, the companies have taken advantage of the design of the tax laws to reduce their tax burden. While PIAC (2013) realises this issue is problematic, it does so only to the extent that the non-payment leads to overestimation of expected oil revenues and hence suggests that the Ministry of Finance and Economic Planning should not consider revenue from corporate taxes when making estimates of oil revenue; not redesigning the tax laws. It follows that, except for royalty payments mentioned earlier in Chapter 3, direct exactions from oil companies are minimal. We shall return to this issue in Chapter 9.

For now, my interest is in examining the neopatrimonial analysis that naturalises these weaknesses and suggests that they are inherent within African public institutions, a view widely propagated by American political scientist Michael Bratton (see Bratton and van de Walle, 1994), the World Bank (1989, 1992), and similarly oriented institutions (see a review by DeGrassi, 2008). It is worth noting the empirical evidence. With the removal of the socialist-leaning Nkrumah government in 1966 and the installation of market-friendly military regimes and a few civilian governments, the democratic and grassroots character defining public life and institutions in Sekondi-Takoradi was systematically undermined as we saw in Chapter 4. First was the delinking of the administration of ports and harbours from railways in 1977 by the Acheampong military regime, ostensibly to enhance the effectiveness of administration but in fact to chop away the radical nature of the railway establishment. Next, entered the infiltration and undermining of the workers' union by so-called 'unionists' and committees of the Provisional National Defence Council (PNDC) formed by Chairman Rawlings, who can be described as a 'Rambo on rampage', to borrow from a popular phrase used by ace Ghanaian investigative journalist, Kweku Baako, masquerading as a revolutionary. Then, since the onset of the structural adjustment programme characterised by a World Bank-endorsed decentralisation model that extolled accountability to market forces and little or no accountability to urban residents, downsizing the state among others through firing public officers. It is these political economic forces – not naturalised Africanised syndrome of corruption and inexperienced public administration – that have contributed substantially to the current compulsive miasma and, in turn, the city authorities have not been successful in self-revival through taxation.

There are major ramifications from the inability to tax windfalls in Sekondi-Takoradi. The transition from a port to an oil city is occurring in an environment of great income inequality, which is fast changing property relations. The power of landlords has greatly increased, emboldening them to recover possession of their property for re-letting to higher income tenants. The *Complaints Book* at the Metropolitan Rent Control office showed that, since 2007, 'recovery of possession', formerly uncommon in the *Complaints Book* now recurrently appears in the records. Furthermore, the share of cases of recovery of possession, inducement to quit, and eviction have become substantial. In 2012, for example, about 52 per cent of the cases reported to the Rent Control Department involved issues of recovery of possession, while 15 and 6 per cent related to inducement to quit and ejection respectively. Interviews with the Police Service showed that some land cases also end up being handled by the service and hence it can be argued that the figures captured in the Complaints Book are only a partial representation of the problem of conflict in the land and housing sector. The future of the majority of indigenes who are not well-to-do in Sekondi-Takoradi is, therefore, bleak, but taxation regulations alone without complementary and substantial changes in broad political economic policy making is unlikely to bring about the future envisaged by advocates of property taxation.

Conclusion

The experiences of Sekondi-Takoradi, where – in African terms – an elaborate property taxation regime is in place calls into question the diagnosis and prescription of tax advocates, but confirms the social dislocation and distress prognosis of failing to tax windfalls. Institutionalising land taxation is necessary but not sufficient to bring about alternative urban development. The *exemptions*, *exceptions*, and broader socio-economic *environment* of the regulatory framework can greatly circumscribe the potency of property taxation as a vehicle for stimulating alternative urban development. Thus, the merits of land taxation, while obvious, are contingent rather than assured. Advocates of land taxation, therefore, ought to broaden their demands, beyond merely establishing property taxation regimes.

These findings make a significant analytical contribution to rent analysis, at least in three ways. First, the view that value increases driven by factors not entirely related to landlords' own investment ought to be creamed off by taxation is incomplete without analysing the capacity of the state to do so. Second, the chapter shows that by not contextualising their claims in the light of neoliberalism, Georgist political economy struggles to understand that neoliberalism drives the state to legislate away its power of land taxation, as consistently argued by David Harvey. Third, there is an urgent need for Georgist political economy to engage a theory of the African state, its administrative and political capacity, and how these features are a product of particular historical and contemporary experiences – as Nwoke and Mahdavy show.

In addition, the chapter makes an important contribution to both the theoretical and empirical literature on taxation and oil production. It demonstrates that the oil industry raises other taxation issues that are not captured by the key concept of 'oil rent', namely the nexus between processes and mechanisms in the oil industry per se and their resulting influences and implications for the surrounding built form in cities. It shows that ways in which the rental dynamics in the built environment are stirred and stimulated by the oil industry in cities are vast with wide social and economic implications and hence ought to be carefully incorporated in the literature to expand and extend our understanding of oil rent into the built environment in cities and communities where oil is extracted. Empirically, it has used the example of Sekondi-Takoradi to show how other studies on the concept of oil rent *with* an urban form can proceed.

To the taxation literature generally, the chapter establishes the need to combine a study of the design of taxes with looking at processes of implementation and enforcement not only synchronically, but also diachronically. In other words, taxation ought to be studied as a system. And, finally, to policy makers, the chapter presents an eye witness account of how the whimsical design and implementation and inattention to taxing oil induced increases in land and real estate markets lead to eviction and hence social exclusion of the poor and meek. So, on all fronts, both theoretically and empirically, we obtain a substantial and original contribution from analysing rent by taking a property rights perspective. What remains to be discussed is how to use the rents and revenues generated.

Notes

1 Under the Australian Petroleum Resource Rent Tax regime, oil rent is essentially profit rent or 'a tax on profits from the sale of petroleum products'. See the full details at Australian Government Taxation Office, www.ato.gov.au/Business/Petroleum-resource-rent-tax/In-detail/What-you-need-to-know/Introduction-to-PRRT/ (accessed 23 August 2013).
2 This is a process by which landed transactions are officially received/recorded by the Land Valuation Division of the Lands Commission.
3 This payment is different from the 'surface rental' which is described in the Petroleum Revenue Management Act, Act 815, section 6. Surface rental is 'fees paid based on the area of licensed held' (Petroleum Revenue and Management Act, s. 61, p. 32).

9 Socialisation of oil rents

Introduction

How can the rents from oil be best socialised? A strategy of nationalisation, as once suggested by Nwoke (1986) can help to end the challenge of sharing oil rents with multinational oil companies and hence will lead to appropriating more rents for social uses. However, current state inadequacies in terms of possessing appropriate oil technology will make any nationalisation plans in the short run abortive. Rather, the government can renegotiate the terms of the agreement to obtain more favourable conditions with respect to appropriating more rent. The high grade quality of Ghana's oil, what is known in the industry as being 'sweet' (see MacCaskie, 2008), should give the state a stronger position in bargaining. In principle, the oil companies can avoid paying more rents to the state by de-territorialising but, given how much they have already invested in the country and in immoveable technology, they are unlikely to relocate. In future, issues about increasing the share of absolute and differential (type 2) going to the state will be desirable, but for the short time, greater effort should be directed towards obtaining more type 1 differential rent. Together with the existing revenues accruing to the state, oil rents can be better socialised.

The rest of the chapter is divided into two sections. The first briefly describes policies of direct cash transfers, education, and public housing as some of the possible ways to utilise oil rents. The second section analyses the preferred government policy on using oil rent for road transport investment and shows the investment undertaken in the road sector to date. The emphasis on road transport is appropriate because it is a key expenditure item for the Government of Ghana, although the law permits investment in other modes of transport. It is, therefore, important to assess carefully the legislative and executive decision to prioritise investment in road transportation. In doing so, the chapter compares road to rail, the latter being one of the modes of transport mentioned in the regulation but, as yet, given very low priority by the state. The analysis is limited to three parameters, namely economy, society, and environment. These are the values which underpin most national policies in Ghana (e.g. Ministry of Local Government and Rural Development, 2010).

The chapter argues that public housing and publicly provided education will be highly beneficial in resolving a worsening housing problem, but direct transfers, while important in giving a 'universal basic income' (Wright, 2010), will be extremely difficult to implement given the lack of records on a vast proportion of the population. Consultation and collaboration with fisher and farming groups and other marginalised groups, however, may help to restrict the income to these groups that have lost their income. Also, while road transport is in vogue, the country would achieve greater benefits – economically, socially, and environmentally – if it invested in railways. Further, the investment in road is likely to worsen social, economic, and environmental pressures for the metropolitan area of STMA. Overall, this chapter defends a combination of public policies on education, public sector housing, and rail transportation, while suggesting that a limited amount of direct cash transfers may be considered for the poorest of the poor.

Direct cash transfer

A widely held view in radical political economy is the idea of universal basic income (UBI). This income is given to all citizens in a country, regardless of income level, whether they are employed or unemployed. Thus, it is not minimum wage, but a guaranteed constitutional right for all citizens. American economic sociologist, Erik Olin Wright, has been a key advocate (see Wright 2010, pp. 4–5 and pp. 217–222). The rationale is that everyone can escape poverty. Its full implementation may lead to all people escaping income poverty, at least. It is also justified on the grounds that it will give all citizens an entitlement to the 'free gifts of nature'. It will have dramatic positive effect on the goal of halving global poverty. A universal basic income challenges the idea that economic growth is a sine qua non for poverty reduction because even for an economy that is not growing, distributing the rents from oil can pull the poor out of poverty. For all these reasons, direct, unconditional cash transfers have been recommended for many oil countries in the world, including Nigeria and Libya (Segal, 2011).

The UBI idea has important drawbacks. First, there is no guarantee that recipients of the money will spend it in ways that will stimulate quality growth of the economy. Second, it is a uniform measure that does not positively discriminate in favour of settlements most devastated by oil. Thus, both the least negatively affected and most negatively affected receive the same amount of money. Overcoming these challenges is not easy. Issues arise about whom to target, how to ensure that the income is used in effective ways, and how to curb administrative problems in the disbursement of the UBI drawn from oil rents (see Genugten, 2011). Even more difficult is the issue about how substantial the amount will be and hence how transformational it will be to the material conditions of people in the metropolis. According to one calculation (Moss and Young, 2009), every adult in Ghana will only receive $80 per annum. This amount is quite small and unlikely to generate substantial multiplier effects especially so when it is not clear how or which structures are in place to ensure that the money is well distributed. Even if the UBI is assumed to be larger than estimated by Moss and

Young, there is also the risk of demand pull inflation and a further concern about the effect of a universal basic income on rental levels in the twin city. More fundamentally, oil is a finite resource and with a growing population, it is doubtful whether the royalties, from which the UBI is calculated, can sustain future generations. From a Georgist perspective, UBI will force up rents in land and hence worsen the distribution of wealth. So, the UBI idea can also be criticised for not dealing with issues of power and inequality and, hence, possibly worsening the wealth divide. Other investment options such as education and public housing are also worth considering.

Education

While it is also well established that with foreign competition or in an increasingly globalised and capitalised world, education alone cannot make the poor rich and solve all the problems of labour, public education is an important strategy to take advantage of the opportunities in the labour market. It is well known that education has huge multiplier effects on industrialisation, job creation, and general economic development. Indeed, the more educated can generally better stand up for their rights (see, for example, Dunn, 2012). So, investing in education is a vital alternative use to which oil rent can be put.

Currently, Ghana has a policy of Free Compulsory Universal Basic Education. However, a lack of state support has led to many hidden costs that, in turn, have led to a situation where education has become highly commodified. The poor do not generally advance in their education. Evidence from the Ghana Living Standards Survey shows that 31 per cent of all adults have never been to school, while a further 17.1 per cent went to school but obtained no qualifications. Thirty-nine per cent of the population only hold the middle or basic education school qualifications, while the remaining 13.6 per cent have secondary or higher level qualifications (Ghana Statistical Service, 2008, p. 8). Success in obtaining education is highly gendered, with males dominating the 'educated class'.

Comparable figures are not available at the Sekondi-Takoradi Metropolitan level. There are indicative figures in the recently published *Western Region Human Development Report* (UNDP, 2013), though. According to the report, which draws on 2010 figures, the net primary school enrolment rates for male and female students respectively are 99.2 and 94 per cent. The net total primary school enrolment rate is 94.4 per cent. Completion rates are very high too. At the primary school level, the report shows that the rates are 114.8 per cent for female students and 101.7 per cent for male students, making a total of 113.9 per cent. The data on junior high school are relatively scanty. Although they show that the net junior high school enrolment rates are 60.7 per cent for male students and 63.7 per cent for female, making a total net junior high school enrolment of 62.2 per cent, they do not shed any light on completion rates.

Further information on education can be gleaned from the *Medium-Term Development Plan* (STMA, 2010) which confirms that education levels in the metropolis are high. For instance, about 66 per cent of the total population of students who sat for the Basic Education Certificate Examination in 2009 passed. Education facilities, enrolment, and completion rates are highest at the basic and junior high school levels. Generally, teachers are mostly well trained, although, like the rest of Ghana, they are poorly paid. The education infrastructure is in dire need of rehabilitation and improvement, as most public schools are old and most do not have toilets and urinals. Library facilities are poor as is the condition of furniture. While the gender gap is small, almost at par, and hence deserving of commendation, the lack of tertiary education facilities beyond the Takoradi Polytechnic is a major problem (STMA, 2012a). Indeed, it suggests that man-power here is trained mostly for vocational and technical roles, leaving aside other high paying roles for university-educated professionals to people outside the metropolis.

A greater investment in quality public education at all levels, but especially senior to tertiary level, will greatly enhance urban economic development in the metropolis. More fundamentally, it will enable locals to be better prepared to assume leadership positions in the oil industry. Given that, in Ghana as a whole, it is school, boarding, and lodging (40.7 per cent) and school fees (28.3 per cent) that take the bulk of education expenditure (Ghana Statistical Service, 2008, p. 13), public intervention in these areas too will have wide impact in reducing the cost of education for parents and guardians.

Oil revenues provide a good opportunity to improve the educational con-ditions in the country. The resources may be used to recruit high calibre teach-ers, stock public libraries, expand educational infrastructure, or improve the conditions of teachers. Some opposition parties in Ghana have recognised this possibility. The Convention People's Party (CPP) and the New Patriotic Party (NPP) went all out to spell out elaborate processes of supporting public educa-tion in the lead up to the 2012 elections. According to the CPP (2012, n.p.), 'We will treat our oil wealth as a "windfall": Revenue accruing from oil will go into investments in infrastructure and 21st century education.' The NPP on the other hand, notes:

> We are fully committed to making secondary education free for every Ghanaian child. By free SHS we mean free tuition, admission, textbook, library, science centre, computer, examination, utilities, boarding and meals. Although the cost of free secondary school education will be high, at an additional 1% of Ghana's total income, the alternative of a largely uneducated and unskilled workforce is a situation Ghana cannot afford. So NPP will prioritise and fund this expenditure using budgetary resources (including resources from oil exports) in the interest of the long-term growth of Ghana's economy.
>
> (NPP, 2012, p. 25)

There was much brouhaha about what 'free' means, prompting the NPP to offer the following explanation:

By free SHS, we mean that in addition to tuition which is already free, there will be no admission fees, free textbooks, no library fees, no science centre fees, no computer lab fees, no examination fees, no utility fees, free boarding and free meals and day students will get a meal for free.

(Akufo-Addo, 2012, n.p.)

Further, the NPP notes, it

involves the building of 350 new Senior High Schools and cluster Senior High Schools to accommodate all the children who finish JHS 18 months after the coming into office of an Akufo-Addo administration. In June 2013, according to the latest Ghana Education Service census, 407,158 students from both 3-yr SHS and 4-yr SHS cohorts will be graduating. This means when Free SHS starts in September 2013 these places will be available. The situation is, therefore, not nearly as dire as the NDC wants everyone to believe.... If the need arises, we will also spend some GH¢25m a year to buy extra school buses for the main purpose of transporting day students to and from schools to make up for the short term problem of insufficient facilities.

(Akufo-Addo, 2012, n.p.)

While the rival NDC was cynical and pessimistic about the feasibility of the free SHS programme, the NPP's own costing put together by a team, including George Yaw Gyan-Baffour,[1] a respected economist in Ghana, pegged the figure at around an additional 1 per cent of Ghana's GDP. If this is the case, the programme is feasible. A major challenge, however, is that the NPP's programme will not be means tested and will not make special arrangements for people in the Western Region. In turn, it may end up serving the nouveau riche, not the poor; and people in more affluent regions and cities. A combination of some revised version of the NPP programme hopefully with inputs from the other parties, the National Development Planning Commission and the Ghana Education Service will do the nation well, as will state provided scholarships to the large cohort of 'needy but brilliant students', marginalised groups such as women, and people from the Western Region. Recent interviews of the growing number of youthful members of parliament (Gyampo, 2013) provide a comforting picture: regardless of political party affiliation, a large majority of the MPs (87 per cent) identify quality education as an area in need of urgent attention by the state. So, there is room to explore how to use oil rents for public education purposes. That said, this support ought to be linked with other social interventions such as public housing

Public housing

Housing is a major problem in urban Ghana, as I discuss in *Governance for Pro-Poor Urban Development: Lessons from Ghana* (Obeng-Odoom, 2013b). It is inadequate in terms of quality and quantity and the rate at which the current deficit is being addressed is far less than the rate at which it is being created. So it is in the Sekondi-Takoradi Metropolis, as described in the *Medium-Term Development Plan* (STMA, 2010). The over 500,000 population are crowded in 36,079 houses in the metropolis with most of them built some 30–50 years ago since when little has been done to them by way of maintenance.

The state could build dormitories and flats for rental purposes. These houses would not be free, of course. They would command 'social rents' below the going market rent. In order to prevent a scenario where these public houses benefit the wealthy more than the poor, rent could be charged as a proportion of the income of the tenants. Thus, in monetary terms, some tenants would pay more rent than others but, in percentage terms, all tenants would pay the same proportion of their income as rent.

It is imperative to ask what the benefits of public rental housing are apart from its relatively low rent. It is well established that public housing has poverty reducing, economic development, and health enhancing characteristics (see, for example, Yeboah, 2005; Arku and Harris, 2005; Grant, 2009; Tibaijuka, 2009; Tipple and Speak, 2009; Njoh, 2012). In general, these beneficial outcomes arise because public housing is provided directly for its use rather than for exchange value. The study of Phibbs and Young (2009, pp. 222–223), for example, has shown that public housing significantly improves employment and health status of occupants and enhances the education of their children. When people spend less on housing, the 'extra' money can go into the purchase of healthy food and good quality healthcare. Public housing is also more secure in terms of tenure as such tenants are not usually arbitrarily ejected. People living in public housing, therefore, have less stress. The evidence from research conducted by Morris (2009) in Australia shows that the quality of life of public housing renters aged 65 and over is higher than that of similar older renters in private rental housing due to lower accommodation costs and greater security of tenure enjoyed by the former.

As with many other public policies, public rental housing has its challenges too. It is based on the assumptions that the state has full information about the incomes of the tenants and that it has the financial resources to provide housing. Both assumptions do not apply in Ghana because of the massive informality and low national income. There is, however, some potential to overcome these problems.

Local governments could help to improve the availability of information on the people in their communities. This information bank would also serve as a source of information for taxation purposes. In this direction, a bottom-up approach that places emphasis on local governments working with the people in their communities would be useful. From working with the local people, local government can then also work for them. There is evidence, which shows that when people take part in formulating the 'social contract' or rules of social life, they are more likely to make the contract or rules work (Ostrom, 1990). That is evidently the case of one peri-urban area in Ghana for which I conducted a separate field study (Obeng-Odoom, 2013c).

As discussed in the previous chapter, taxation as a source of revenue is also important. The use of land tax or a tax on undeveloped land could have particular benefits for housing provision, but these houses would be insufficient. Oil rents can be another source of revenue. Assuming that the country will earn about US$836 million annually from oil production and about half of this sum will go to the country without any significant costs being incurred (Breisinger *et al.*, 2010), Table 9.1 shows some possible scenarios of oil-financed affordable housing over a ten-year period. Breisinger *et al.* (2010) figure ($418 million) is a conservative average, if we make a further assumption that the receipts will increase over time (cf. the 2012 total receipts of US$541.07 million). However, it gives us a sense of what can be done with the receipts from oil revenues.

Table 9.1 shows that if half of the expected revenues from oil were to be invested in building affordable houses, over 200,000 houses could be built in a period of ten years. Together with the housing construction financed from land tax, over 300,000 houses nationwide within ten years could be expected. Land values, as we saw in Chapter 5, are clearly higher than the US$10,000 used in the table. However, drawing on the insights of Henry George considered in Chapter 2, the assumption here is built on the expectation that betterment charges and taxes on land will force down land values as landowners are made to pay for 'unearned incomes' and estate agents are discouraged from speculating. However, the valuation is done, whether by the cost approach or the income capitalisation approach, land values must fall. If it is by the cost approach, with falling land values, overall capital values should fall. Using the income approach, overall capital value and indeed land value should fall too because the full rental value, even if the capitalisation rate remains the same, has fallen. I have done a more careful commentary on valuation methods elsewhere (2012b). In my book, *Governance for Pro-Poor Urban Development* (Obeng-Odoom, 2013b), pp. 108–124), I go into details on how to attain such low housing prices. For the current book, it will suffice to say that the assumptions of land values are based

Table 9.1 Possible scenarios of oil-financed affordable housing

% of petroleum receipts set aside for housing	Equivalent amount (000s)	Number of houses assuming 1 house costs US$10,000	Total number of houses in 10 years
5	20,900	2,090	20,900
10	41,800	4,180	41,800
15	62,700	6,270	62,700
20	83,600	8,360	83,600
25	104,500	10,450	104,500
30	125,400	12,540	125,400
35	146,300	14,630	146,300
40	167,200	16,720	167,200
45	188,100	18,810	188,100
50	209,000	20,900	209,000

on conceptual groundings in our theoretical framework, established valuation principles, and empirical research in Ghana.

The social rent from these houses could be reinvested in providing more houses if there is an economic surplus over the costs of maintaining the existing ones. It could be argued that tenants may default in paying their rent. The eminent housing expert Abrams had to deal with this concern when he recommended the Roof Loan Scheme to the Nkrumah government in 1954. One minister told Abrams' team that 'our people look on the Government as their father, and you don't repay your father' (Taper, 1980, p. 52). The claim by the Minister may have been correct but, following Ostrom (1990), one could argue that such apathetic behaviour is more likely to emerge if tenants do not participate in decisions related to their housing. Also, the work of Obeng-Odoom (2009) shows that though perception plays a part in default rates, a more important cause of default is poverty. These considerations mean that if tenants are offered sustainable employment, the rate at which they would default on their rent is likely to be low. Public housing itself could stimulate a desire to work because it relieves tenants of the problems of dealing with private landlords (Glynn, 2009) and can be part of a bigger project of empowering people.

The suggested social policies of direct cash transfer, education, and public housing are all plausible, but they are currently not a major focus of attention for the Ghanaian state, which is currently prioritising oil investment in road transport. It is important, therefore, to scrutinise this aspect of policy. To do so, the rest of the chapter considers how road transport performs against another transport alternative – rail. Data on road transportation for this analysis were gathered from the Ghana Highway Authority, Driver Vehicle and Licensing Authority, Ghana Tourism Authority, and Ahantaman Rural Bank. For the rail sector, data were collected at the traffic control department of the Ghana Railway Company, and historical material from in-depth oral interviews with four railway workers who have been working for the sector since the 1970s, have risen through the ranks, have held top positions in the railway workers' union, and hence have developed intimate knowledge of the sector. Finally, I gathered data informally from taxi and *trotro* (small minibuses) drivers and did many transect walks in the city in a three-month field research, spanning December 2012 to March 2013 mainly based in Sekondi-Takoradi – Ghana's oil city; but also travelling and researching throughout the Western Region – the oil region, as explained in Chapter 2.

Oil rents and transport investment

Ghana has a massive infrastructural deficit, so the country's parliament has legislated that oil rent should be channelled into infrastructural development. Regarding transport, the Petroleum Revenue Management Act, 851, notes that where there is no national development plan with parliamentary approval, as is the case in Ghana now, the revenue from oil (annual budget fund) should be used for 'infrastructure development in telecommunication, road, rail and port' (section 21, subsection 3d). However, the current government, on the basis of priorities set out in the *Ghana Shared Growth and Development Agenda, 2010–2013* (Government of Ghana, 2010a), has prioritised road transport investment (Public Interest and Accountability Committee [PIAC], 2012). According to the growth agenda (Government of Ghana, 2010a, p. 69), the Government of Ghana should:

> prioritize the maintenance of existing road infrastructure to reduce Vehicle Operating Costs (VOC) and future rehabilitation costs; improve accessibility by determining key centres of population, production and tourism; re-instate labour-based methods of road construction and maintenance to improve rural roads and maximize employment opportunities.

For this reason, the government has consistently spent oil revenue on road maintenance and construction. In the year 2011, some 80 per cent of oil revenue was spent on road infrastructure (PIAC, 2012). A vivid description of the investment plan was given to the Parliament of Ghana when the Minister presented the 2011 budget. According to the Petroleum Revenue Management Act, '[t]he programme shall be reviewed every three years after the initial prioritisation, except that in the event of a national disaster, the Minister may make a special request to Parliament for a release of revenue' (section 21, subsection 6). The 2012 Budget (Government of Ghana, 2012) shows and PIAC (2013) confirms that the Minister complied with this regulation; meaning investment in roads remained a key priority area taking a disproportionate amount of oil rent-related investments.

The bias for road investment is not new in Ghana's political economy. While Ghana's *Coordinated Programme for Economic and Social Development Policies, 2010–2016* (Government of Ghana, 2010b) identifies multiple modes of transport in Ghana as in need of investment, the document gives road transport first mention and priority and only mentions others, including rail, as auxiliary. In its own words,

> [t]he entire railway rehabilitation, modernisation and expansion programme is aimed at reducing the increasing pressure on urban transportation in the major metropolitan areas of Accra, Kumasi, Tema and Sekondi-Takoradi and provide freight haulage to the new oil and gas-driven industries anticipated to spearhead accelerated growth. The expected outcome will be the reduction of pressure and congestion on the roads and highways.
>
> (Government of Ghana, 2010b, p. 56)

Indeed, rail appears to be consigned the colonial role of only helping to exploit mineral resources: '[t]he existing railway network will be rehabilitated or totally re-built as appropriate, modernized and expanded to support accelerated *industrial* growth' (Government of Ghana, 2010b, p. 55, emphasis added). Table 9.2 shows that the rail sector has suffered years of neglect.

Table 9.2 State of rail lines in Ghana, 2000–2011

Year	Track length (km)	% in operation	Route length (km)
2000	1,300	90.2	947
2001	1,300	90.2	947
2002	1,300	64.7	947
2003	1,300	43.65	947
2004	1,300	43.65	947
2005	1,300	43.65	947
2006	1,300	44.7	947
2007	1,300	46.65	947
2008	1,300	35.63	947
2009	1,300	35.63	947
2010	1,300	36.09	947
2011	1,300	33.89	947

Source: field data (GRC Transport Indicators Database Project, 2012/2013).

Both the track and route lengths have been stagnant at a pale 1,300 km and 947 km respectively, making the combined track and route length as of 2011 22.25 times less than the total road length of 49,996 km in the 2000s (Addo, 2006, p. 17). Further, the share of the short length of track in existence that is operational has been declining over the years. The disinterest in the rail sector also plays out in terms of the employment profile of the sector which can be seen in Table 9.3.

Between 2000 and 2011, employment consistently declined. The strength of the railways workforce has more than halved in just about 11 years, as the government has retrenched some members of staff, while others have retired. In all these, successive governments have decided not to recruit new people, leading to a depletion of the workforce. What is not shown in Table 9.3, but which became evident through my interviews with the railway workers, is that the workers also

Table 9.3 Number of employees of the Ghana Railway Company, 2000–2011

Year	Number of employees
2001	4,318
2002	4,224
2003	4,092
2004	3,843
2005	3,635
2006	3,308
2007	2,872
2008	2,388
2009	2,370
2010	2,381
2011	2,273
2012	2,154

Source: field data (GRC Transport Indicators Database Project, 2012/2013).

suffer delayed payment of their meagre salaries. From the foregoing, it is imperative to consider the relative importance of road transport in the light of the disproportionate attention it receives from the Government of Ghana. To do so, the next section uses a common basis of comparison, namely contribution to economic, social, and environmental development, which are the key aspirations of national policies in Ghana (see Government of Ghana, 2010a, 2010b).

Road and rail transport compared: issues in economic, social, and environmental efficiency

Investment in road infrastructure occupies the uncontested priority position of the Government of Ghana. The advantages of building more roads and rehabilitating old ones have been so frequently stressed that it becomes repetitive to describe them here. S.T. Addo, one of Ghana's foremost writers on transport, has consistently written and spoken about these issues (see, for example, Addo, 2006). Indeed, Addo argues that transport generally, but road transport especially, is seen as 'development' and modernisation in the country (2006). Coupled with these representations is the ostentatious value of cars (Chalfin, 2008). So persistently are the advantages of road transport touted that the cost of expanding road infrastructure is often overlooked. Yet, the achievements in road construction and hence road transportation have come at great social and environmental cost to the country. It is one of the most common causes of death in Ghana. Since 1991, road accidents have claimed the lives of over 32,000 people and injured over 256,000 people (Building and Road Research Institute [BRRI], 2012, p. 33).

The diagnosis of the causes of road accidents differs widely. For some local people, traditionalists, and religious groups, witches and spirits are the primary cause of road accidents in Ghana and hence more rituals are needed to exorcise the demons on the roads (Klaeger, 2013). On its part, the government claims that few and poor quality roads constitutes the primary reason for accidents, and hence there is the need to build more and better roads and maintain old ones. The prevalence of poor quality vehicles has also been identified as a major cause of road accidents (Poku-Boansi and Adarkwa, 2011). More recently, the Building and Road Research Institute has revealed that it is the role of drivers that contributes the most to road accidents. In 2011, for example, 89.4 per cent of accidents leading to injuries and fatalities were attributed to driver errors such as over speeding and inattention (BRRI, 2012, p. 25).

Road transport is private business in Ghana and there is hardly any intra-city public road transport. The *trotros* are erroneously called 'public transport', but these are private investment vehicles of individuals, some of whom are quite or very wealthy. In their haste to reap profit, most car owners tend to employ unprofessional and hence inexpensive vehicle drivers. One study found that 95 per cent of commercial drivers in Ghana possess no licence to drive (ABLIN Consult, 2008). Further, most car owners tend to import second-hand cars, and pile pressure on drivers to meet high daily targets. In turn, most drivers tend to wrongly overtake, tend to overwork, and tend to over speed – all the time worrying about how to meet the high targets set by their task masters. Partly as a product of this cycle, drivers also tend to bribe regulators who try to enforce road safety rules. In the end, accidents proliferate (Obeng-Odoom, 2013b, pp. 89–107). So, road accidents are systemic.

These dynamics are at play in the oil region (Western Region) within which is the twin oil city of Sekondi-Takoradi. Between 1991 and 2011, there were 15,080 accidents from which there were 22,340 traffic casualties. Table 9.4 provides the equivalent figures for Sekondi-Takoradi.

It shows an increasing trend across all categories: accidents, casualties, and fatalities. While the two large cities, Accra and Kumasi, have a worse accident profile (see BRRI, 2012), the discovery of oil in Sekondi-Takoradi is likely to make it a dangerous city because of five reasons gathered from taxi and *trotro* drivers, the Head of Credit (including credit for transport) at Ahantaman Rural Bank, and the Head of the Ghana Tourism Authority in the Western Region. First, more people have become interested in the road transport business, surmising that the increase in the population of the metropolis creates a market for transportation services. Second, there is likely to be a greater use of taxis as inhabitants become wealthier. Third, as we saw in Chapter 5, the industry is making some people richer and some can afford to become car owners who employ drivers and force them into risky behaviours that increase the likelihood of accidents. A fourth reason for the possibility of more cars and greater vulnerability to accidents is that the number of car rentals or the number of people in car renting business has increased dramatically and with competition comes the now familiar cycle of accidents in the country. Finally, the idea of transport micro credit in which private banks are giving loans to people to go into the car business is likely to elevate competition in the road transport business to new levels. The banks show a drive for profit, but also claim that they would want to help people who require employment. This financial scheme has been well received, with the insignia of various banks on commercial vehicles in the city announcing their affiliation but also inviting others to seek car loans to enter the commercial and private car business.

A few drivers have tried to game this transport microfinance scheme by obtaining loans from multiple banks to purchase different vehicles. Quite apart

Table 9.4 Accidents, casualties, and fatalities in Sekondi-Takoradi, 2001–2011

Year	Accidents	Casualties	Fatalities (% of casualties)
2001	181	171	9.94
2005	174	152	12.50
2010	185	182	13.74
2011	140	118	13.56
Total	680	623	

Source: compiled from Building and Road Research Institute, 2012, p. 85.

from the obvious outcome that this practice can lead to an explosion in car population in the metropolis, there is also increasing pressure on drivers to meet their targets and amortise the loan they have taken. Added to this pressure is the issue of frequent car breakdowns – a not too surprising phenomenon given that most of the cars purchased with loans are second hand. Or, pressure can increase when, at a point, there are so many cars such that the system is saturated and hence the market for individual taxis has shrunk. To survive, some drivers have started driving long hours, working overtime, over speeding, and overworking. And, as we have seen, these features tend to lead to more and more accidents.

In addition to the poor socio-economic record, the road sector has a degrading effect on the environment. Alone, it contributed 4.6 million tonnes of carbon dioxide equivalent in 2005 and, unless something fundamentally different is done, a 36 per cent increase is predicted (Faah, 2008). This trend has dire consequences for coastal cities like Sekondi-Takoradi through erratic but torrential rain and possible flooding (Obeng-Odoom, 2013b, pp. 89–107). While other factors contribute to emissions, such as the type of fuel used, in Ghana, the principal reason is the age profile of imported vehicles. Only 8 per cent of imported cars are brand new. The rest can be anything from second, third, fourth, or even fifth hand (Yeboah, 2000; Linder, 2006).

While the use of newer cars which work with gas instead of petrol contributed to some 9 per cent reduction in emissions between 2000 and 2006 (Environmental Protection Agency, 2011a, p. 54), most commercial cars in Ghana do not use gas, mainly because they are old models. Additionally, the existing regulation does little to deal decisively with emissions because it places a penalty only when a very old car (ten years and over) is imported. In turn, old cars aged under ten years can be imported without penalty. Even with penalty, it is not clear how deterrent is the policy. In turn, the rate at which cars – including old ones – are imported into the country far outweighs the rate at which the population increases (Kayoke, 2004; Obeng-Odoom, 2013b, pp. 89–107) – setting in motion a vicious cycle: more cars, more congestion, more emission.

Rail, however, shows a robust contribution to the economy, and positive influences on society and environment. Indeed, the country developed economically largely on the back of the forward linkages between the rail sector and the economy, expressed in lower transport cost, commodity extraction, and exportation through rail (Tsey, 1986, 2013). Historically, rail transport has been a major contributor to Ghana's development, economically and politically. Harvard historian and leading Africanist, Emmanuel Akyeampong, commenting on the rail sector in its booming years describes customs duties and rail revenues as the 'two leading sources of government revenue' (1994, p. 397). The contribution of rail revenue to the government coffers consistently increased between 1910 and 1913: 24.7 (1910), 26.5 (1911), 26.8 (1912), and 27.5 per cent in 1913 (Akyeampong, 1994).

The role of rail in the development of commodities such as manganese, bauxite, timber, and cocoa, and the development of cities and municipal services has been well documented (see Tsey, 1986, 2013), so only a selective sketch is provided here. The railways help in carting goods, and transporting passengers. In addition, their stations were a great meeting point and a venue for the exchange of pleasantries, pleasantness, and ideas. Inscriptions such as 'Food and Local Drinks Base', 'Railway Catering Service', and 'Brighten the corner where you are' remain, but their actual vibrant and happy aura are long gone. Rail transport was instrumental in opening up new areas for cocoa production, instrumental in founding new settlements such as Kumasi, and instrumental in promoting national integration (Korboe and Tipple, 1995; Tsey, 1986, 2013). Rail alone employed 62.6 per cent of the workers in the formal sector in Sekondi-Takoradi, the headquarters of the service (Busia, 1950), and the workers constituted such a well-organised group that they became the most influential workers' union in the Gold Coast, as Ghana used to be called (Jeffries, 1978). Rail service freed labour from doing menial jobs of head carriage to doing other economic activities. Indeed, it expedited the distribution of goods and services which hitherto took long hours to accomplish. The view of one observer was that 'a railway train of average capacity will do the work of 13,000 carriers at one-twentieth the cost' (Tsey, 1986, p. 288). The railway sector propelled both internal and external trade, leading to great social and economic advancement, even if, because of the historical times, imperial interests benefitted disproportionately from the developments (see Tsey, 1986, 2013).

With its great influence also came political significance, as the railways workers used their numbers and organisational skills to help in the independence struggles and to force the governments of the day to improve the conditions of labour (Busia, 1950; Jeffries, 1978). There is even a bigger, unexplored potential for the rail sector: its backward linkage. Historically, only very little inputs from the local economy have fed the rail sector. Part of the reason was that the local economy was organically linked to the coloniser's economy in Britain, but it was also the case that supporting local industries were yet to be developed (Tsey, 1986, 2013). It follows that, in independent Ghana, with over 50 years of political maturity and economic development expressed in the growth of many local industries – both formal and informal – if an indigenised version of the inputs to the rail sector is developed, the contribution of the sector will substantially expand.

Unlike road, rail has little detrimental socio-ecological footprint. Rail accidents are relatively few, as are injuries and fatalities. Table 9.5 gives a more detailed picture of how rail performs in terms of passenger traffic, accidents, injuries, and fatalities.

Although the rail sector has been operating below its capacity, the data in Table 9.5 show that it has a low record of accidents. More fundamentally, human error as a cause of accident is marginal in the rail service where accidents (derailment) relate to underinvestment in the sector. Even then, the accidents are few, compared to the road sector and the negative consequences (injuries and fatalities) considerably less. The train drivers are qualified, take regular rests, and are in no hurry to make extra money or meet set targets – being public, salaried workers (Tsey, 1986, 2013).

Table 9.5 Passengers, accidents, and consequences in Ghana, 1990–2011

Year	Number of passengers	Number of derailments	Number of collisions	Number of persons injured	Number of fatalities (persons)
1990	2,228	92	N/A	N/A	N/A
1991	1,305	94	N/A	N/A	N/A
1992	1,270	79	N/A	N/A	N/A
1993	1,459	91	N/A	N/A	N/A
1994	2,300	62	N/A	N/A	N/A
1995	2,208	68	N/A	N/A	N/A
1996	2,077	77	N/A	N/A	N/A
1997	2,105	122	N/A	N/A	N/A
1998	2,208	122	N/A	N/A	N/A
1999	1,469	152	N/A	N/A	N/A
2000	844	158	N/A	N/A	N/A
2001	546	151	N/A	N/A	N/A
2002	1,543	157	N/A	N/A	N/A
2003	2,335	135	N/A	N/A	N/A
2004	2,564	151	N/A	N/A	N/A
2005	2,134	136	N/A	N/A	N/A
2006	1,458	113	N/A	N/A	N/A
2007	985	81	N/A	N/A	N/A
2008	950	30	8	0	0
2009	1,120	37	2	0	0
2010	1,377	66	6	0	0
2011	1,320	38	8	1	0

Source: field data (GRC Transport Indicators Database Project, 2012/2013).

Note
N/A = Not available/No available data.

Rail transport produces less emission, according to the Environmental Protection Agency of Ghana (EPA) which found that rail only burnt a pale 0.3 per cent of total fuel consumption in Ghana between 2000 and 2004, whereas road transport used up 93 per cent of fuel around the same time. The transport sector overall burns considerably larger amounts of gasoline, but almost all of that is ascribed to the road sector. For example, in the year 2000 – being the year for which statistical information is available – the road sector consumed 99.3 per cent of gasoline, while rail consumed only 0.6 per cent (EPA, 2011b, p. 53). Of course, if more rails are used, the percentage consumption of fuel will rise but there is no evidence that it will do so, on a per capita basis, at the same rate as road. In terms of cost too, the railway is cheaper. It has been estimated that for the same load, the road sector carries 'at an additional cost of $1 per tonne for the manganese ore and more for bauxite' (Foster and Pushak, 2011, p. 12).

Thus, in terms of society, economy, and environment, the rail sector performs much better than the roads sector. Further, continuing investment in road is unlikely to augur well for the environment and society because of its tendency to pollute, to inure, to damage, and to kill via accidents. Investment in the roads sector merely socialises capital accumulation, that is, it leads to a situation where private profit is increased using public oil funds to build roads. In turn, the public funds, although used for public purpose do not necessarily translate into outcomes for the public good. Investment in rail is more likely to lead to higher and inclusive economic development: greater economic growth, greater distribution of resources, and greener growth. Rail is safe and cheap; therefore, it is likely to be patronised as it was before underinvestment kicked in.

So, if investment in rail is likely to produce better economic, social, and environmental outcomes, why has the Government of Ghana neglected the sector? In the next section, I draw on oral histories from interviews with the railway workers and empirical studies (e.g. Jeffries, 1978; Haynes, 1991) to argue that the reasons are both historical and contemporary. On the one hand, successive governments have shown a morbid fear of the rail sector, its radical unionism, and opposition to governments in power and, on the other hand, governments have been wary not to offend the ideology of private-sector led development in which the country's development partners and some of the governments themselves believe.

'Logo the anchor'

Historically, the rail sector has been a source of morbid fear by politicians and policy makers. The unionisation of railway workers, their radicalism, and activism have been a source of great worry to governments over the years. In his seminal study of the railway workers of Sekondi, Richard Jeffries (1978) documented the solidarity among the railway workers and pressure they exerted on the government of the day, and how, by their war chant of 'logo the anchor' – a declaration that they were striking – economic activities were severely affected. Historically, the railways and the ports were coupled. So, when there was a strike, passengers could not move, neither could freight, by sea or by rail. The economy, dependent on mineral resources and raw materials like cocoa would similarly ground to a halt. The politicians of the day struggled to deal with the railway workers who constituted the largest and most militant labour union. While the union had its own internal wrangling, as is the case in most human institutions, it was united in opposition to the system most of the time. It opposed colonial rule and opposed many ruling governments too.

Therefore, successive regimes always plotted to destabilise the railway workers. Dictator Colonel Kutu Acheampong was perhaps the first leader of the country to deeply undermine and break up the union. He succeeded in decoupling the port from the railway organisation, promising that it was that policy which was going to propel the railway service. There was some opposition from some of the union leaders, but it was not sustained and Acheampong seemed to have convinced more than ruffled feathers. So, in 1977 there was a decoupling of the railways from the ports and harbours. That singular action brought the railways close to its knees, as the sector started feeling the pressure of limited funds and a major attrition in the force of mass mobilisation. The historic split made the railways service totter, giving meaning to the maxim: divided we fall; united we stand.

As we saw in Chapter 4, however, the nail in the sector's coffin came from a familiar source: the World Bank through its Rehabilitation Programme implemented in 1987 as part of Ghana's Economic Recovery Programme of the withdrawal of government or the extension of it to strengthen markets. Indeed, this view represented the broader World Bank position as captured in the 1994 World Development Report, *Infrastructure for Development* (World Bank, 1994). In that report, the Bank took the view that the comprehensive support of the Government of Ghana to the railway sector, culminating in 60 per cent of railway revenue as government subsidies, was a clear case of 'financial inefficiency and fiscal drain' and hence called for the reform of the sector (p. 29). The Bank promised that the railway service would perform much better with limited state involvement.

The Government of Ghana toed the ideological lines of the Bank and has been withdrawing its support of rail transport over the years. The Provisional National Defence Council (PNDC) led the process of state withdrawal, in the hope that the performance of the sector would dramatically improve as promised. Yet, as shown in Table 9.6, the policy of withdrawal has not succeeded in rescuing the rail sector. Rather, its performance has been declining over the years. While the railway organisation was issued with a Certification of Incorporation on 7 March 2001 with the intention of attracting more private sector interest, the company has consistently seen a decline in almost all aspects of its operations – commodity transport, passenger service, and labour strength.

No private investor has found the railway a profitable enterprise. So, after decades of the Bank's advocacy of private sector involvement, World Bank researchers Vivien Foster and Nataliya Pushak (2011) have finally admitted that concessioneering is unlikely to salvage the rail sector. The railway workers have been exonerated. While the workers agree that frequent strikes and lackadaisical work attitude by a few people in their ranks have not augured well for the service, it is tied up with more general problems plaguing the sector. For instance, the workers' emoluments and general benefits have been deteriorating badly, leading some of them to resort to pilfering and other unproductive attitudes (Haynes, 1991).

Table 9.6 Commodity and passenger traffic of Ghana Railways, 1965–2011 (000s)

Year	Cocoa	Timber	Bauxite	Manganese	Total	Passengers
1965	419.33	546.46	276.75	597.65	2,288.12	7,798
1975	114.02	150.70	377.98	317.97	1,191.13	6,934
1985	33.30	35.00	134.50	269.00	509.89	2,104
1995	16.91	55.54	518.11	187.55	813.17	2,208
2005	22.02	36.53	526.45	1,203.97	1,826.57	2,134
2011	–	–	20.94	615.72	636.93	1,320

Source: field data (GRC Transport Indicators Database Project, 2012/2013).

The hardships unleashed by the Economic Recovery Programme of the World Bank and the IMF aggravating deteriorating conditions of the workers brought matters to a head when a major strike was organised in 1992 on the eve of Ghana's transition to the fourth republican constitutional democracy. In a well calculated 'attack' planned at 'Bottom Tree' – the union's spiritual home – the railway workers began assembling their wagons which, as a result of underinvestment, were rickety, in preparation for a surprise 'visit' to the Castle – the seat of presidency in Ghana until 2013. The security agencies picked the signals that the workers were about to 'visit' the government in Accra and quickly rounded up the leaders. However, the security agents were outwitted, as the leaders feigned ignorance of the events and even asked to be given walkie talkie devices for them to help the security agencies in bringing the railway workers under control. These requests were granted, but instead, the leadership used the walkie talkies to enhance co-ordination for the demonstration, the famous visit. The plan was to use several wagons and several routes as a risk diversification strategy.

The journey to Accra was done by different wagons, therefore, while the leadership travelled by road – again a risk diversification strategy. The leaders were arrested midway at Weija Junction by national security, including the Inspector General of Police at the time. However, the explanation by the leadership that it was on its way to stop the railway workers – totalling about 12,000 people – at Kotoku Junction, secured their release. Apparently, the security believed that the leadership would co-operate with them and, given that the leadership had claimed that the number of workers was in the region of 12,000 people, the security was visibly apprehensive. Unfortunately for the national security, the leadership had more solidarity with the railway workers than the state. So, the leaders of the union joined their colleagues who had arrived in Accra earlier – around five in the morning, and were warming themselves up by chanting their war song, *yerekoo yerekoo … yerekoo Ministry* (to wit, we are going, we are going, we are going to the Ministry).

They successfully arrived in the Ministries area in Accra, and took the opportunity to look for the Minister of Finance who had then been whisked away to safety. He later met the leadership, but he was not very sympathetic, surely not when the workers had vandalised state property in the Ministry and lowered the national flag of Ghana, while hoisting their own *asafo* railway flag! The workers, who by this time were acting no longer under the instructions of their leaders, were even more charged and fired up after their leaders' encounter with a not-so-sympathetic Minister and they decided to head for the Castle – the seat of the presidency at the time – although they had no idea how to get there.

Yet, after some wrong turns (including ending up in the power selling company (VRA premises), the state unleashed its full security apparatus to crush the 'insurrection'. The railway workers retreated, but not without a show. The police had attacked on horses and, initially the railwaymen had taken to their heels. However, in the process, one railway worker, who had had some experiences with managing such animals, advised his colleagues not to run but to show their red bands to the horses. Apparently, the colour red is abhorred by horses and it turned out that the strategy brought the railway workers temporary reprieve. But, their victory song did not last for long, as the police and the army who had called for reinforcement attacked with batons and looked fierce with their full armoury. Some railwaymen got injured. In turn, the workers retreated and congregated in the train station in Accra where at the direction of the leadership they entrained to Sekondi-Takoradi, their base to cast their votes and allow their thumb to speak for them in the elections that were scheduled for the next day.

Since the historic 1992 demonstration, no major demonstration of that scale has been organised. A number of reasons has been given for the declining interest in demonstrations, ranging from reduction in workforce, through fear of being fired at a time when there is a high rate of unemployment, and a genuine interest in trying to salvage the railway service (Haynes, 1991). From 1992 to 2000, Jerry Rawlings led the country, and from 2001–2008, John Kufour did. While in some ways Rawlings' charisma and popularity endeared him to the railway workers, he did little to rescue a sinking sector. Indeed, it was in Rawlings' time that neoliberalism took firm roots and shattered the rail sector (see Haynes, 1991, for Rawlings' regime's engagement with the railway workers). Like Rawlings, John Agyekum Kufour made several apparently concerted attempts to improve rail, including the establishment of a Ministry of Railways and Ports (2001–2008). However, these efforts have remained tokenist, whimsical, and disjointed. So, on one occasion, when the sector minister in John Kufour's government met the railway workers and was, as usual, making them promises, workers thinking they were mere empty promises invoked metaphysical powers to punish the Minister if he failed to honour his promises – much to the displeasure of the Minister. That Ministry was dissolved when the John Mills government assumed the reins of power from 2009–2012.

Since 2009, there have been promises and little action by the National Democratic Congress government initially led by John Mills and now by John Mahama (2013–). Overall, no leader, party, or government seems to be too keen to revive a sector well known to breed 'stubborn' workers. Instead, all these regimes appear to be more comfortable to let the market take care of the railway sector – a position which is consistent with World Bank neoliberal ideology.

Conclusion

This chapter has considered various ways of socialising oil rents. It has looked at possible uses of oil rents, ranging from direct cash transfers, public investment in education, and housing; and showing why the last three hold more promise than direct cash transfers while at the same time endorsing direct cash transfers for the poorest of the poor.

The bulk of the chapter, however, has tried to analyse the current utilisation of the rents from oil for road construction, juxtaposing it with rail, and examining the two in terms of their contribution to economy, society, and environment. It has demonstrated that on all these criteria, investment in rail transport seems to be a better way to utilise oil rent going into the transport sector. Road investment amounts to a socialisation of the capital accumulation process, and gives a disproportionate amount of the public oil rents to a few owners of private vehicles. Yet, most of these private commercial vehicle owners, in their bid to earn more profit, tend to employ unqualified drivers for whom high targets are fixed. Consequently, the drivers over speed, overwork, are overtaken, and wrongly overtake, all of which culminate in high numbers of accidents and casualties. Deaths are a major drain on the national and local economy through loss of human capital. Further, road transportation contributes substantially to increasing emissions in the country, through congestion and the burning of large quantities of fuel, making Ghana increasingly prone to climate change.

Rail transport, on the other hand, is cleaner and safer. It emits less, is less involved in accidents, and, in its vibrant days, was a major contributor to economic development. It may take a while and more funding for the service to be restored to its former, booming condition, but its long-term effect is positive overall. Spending on rail will be utilising the public money for the public purpose and for public good. Reasons such as the ideological pressure by the world development institutions and the internal historical fear of the railway sector as rascally and an opponent of the government are likely to inhibit reform – especially when the state continues to exhibit neoliberal, rentier, and comprador characteristics to borrow respectively from Harvey, Mahdavy, and Nwoke. However, civil society presence can help to turn attention to rail too. Civil society groups can exert pressure on the government in terms of giving attention to rail. Currently, however, the attention of civil society and media has mainly been on accounting for oil rent not how oil rent is spent, as we have seen in previous chapters. There is some hope that this might change in the future, especially when the government has taken a US$3 billion loan from China, some of which, it claims, will be used to revamp the railway sector as we saw in Chapter 3.

As this chapter has shown, however, a policy mix that tries to extract more rents and uses a combination of direct cash transfers (limited amount), public education, public housing, and public railway development will clearly lead to better and wider societal impact of oil for the marginalised in Sekondi-Takoradi and the remaining urban centres. Certainly, oil development will have great impact on regional and inter-regional transportation and help to enhance urban mobility, especially if an integrated view is taken emphasising public housing that is served by rail and a school system in which schools are reached by rail services. The presence of railways also tends to drive up land values and, at least from a Georgist perspective, these windfalls can be captured as we saw in Chapters 7 and 8 for further investment in social services as this chapter has shown. A reflection on the challenges, prospects, and lessons from Chapters 1–9 is now what is needed in the story of petroleum and urban development in Sekondi-Takoradi and it is to such a consideration that the book finally turns.

Note

1 Prof. Gyan-Baffour was instrumental in explaining the costing of the NPP to the public.

10 Sekondi-Takoradi

Challenges, prospects, and lessons

The complexities of the oil experiences of Sekondi-Takoradi confound existing representations of expectations and the role of oil in Africa's political economy. Evaluated against the evidence marshalled in this book and interpreted using ideas from Henry George, David Harvey, Hossein Mahdavy, and Chibuzo Nwoke, both the old orthodoxy that oil is a blessing and the new orthodoxy that oil is a curse are clearly misleading, simplistic, and partial. Also partial is the view about the state exhibiting a disembodied 'culture of corruption', even in its rentier form.

There are some economists of 'free market' inclination who claim to reject the resource curse thesis, but then contend that oil in a state dominated economy is a problem. That is, oil produces severe curses when the state is in charge of oil management, so the private sector, private property rights, and markets ought to be extended. For these economists, the policy initiative is to privatise oil contracts and remove the state as far as possible from oil management. In other words, the more markets, the better. This laissez-faire view is ideologically biased and insensitive to socio-economic history, social structures, and the dynamics of institutions and how they are connected to the world system. Being ahistorical and class blind is not helpful because grasping the social foundations and dynamics of capitalism, both temporally and spatially, is crucially important to social analysis and public policy making.

It is more useful to investigate how we got here, the drivers of change, problems inherent in the process, and how these are distributed according to social structures, institutions, political and economic position in the capitalist order; consider different visions of change, and study the prospects and barriers in the journey towards that vision; hopefully one of a good city.

This book has contributed to this particular aspect of Ghana's and hence Africa's political economy, drawing on evidence collected, interrogated, systematised, and synthesised over a four year period. The evidence is clearly that there is massive investment in Sekondi-Takoradi by myriad actors: central government, local government or urban authorities, international oil companies, local entrepreneurs, and foreign capitalists, traditional authorities, workers, and the partners of some oil workers. There is a boom in direct and indirect employment in the metropolis, rendering commentaries about oil generating an 'enclave economy' incomplete at best. In turn, there has been a major interest in the twin city, as workers and investors have migrated into it. Accompanying these changes have been substantial increases in land and housing values, widespread dollarisation and commodification of the land sector, dynamics which further draw investors into the metropolitan economy or whet the appetite of those already in the twin city to expand capital circulation. Petty investors and speculators such as some real estate agents have poured into the twin city or have been drawn from other sectors into the property market or simply have expanded their work to include real estate investment. This process of the creation of wealth ought to be viewed critically for it reveals that constructed and imposed expectations are exaggerated, and the outcomes of the wealth creation process concentrated, rather than spread.

The boom in job creation is significantly below expectations either on the rigs or off the rigs. The result is massive disappointment by the youth but also others who had high expectations and some of whom therefore went to obtain various oil-related training. Even worse, job creation has come with massive job losses and, in turn, will require further evaluation of the net effects of job creation in Sekondi-Takoradi during its oil age. Furthermore, the jobs created on the rigs have been those concentrated on the lower rungs – a feature that is connected with the general maldistribution of the benefits and losses attending the oil industry and calls into question the optimistic claims by the state in its local content policy.

Already wealthy people have cashed in on the investments. Elites such as chiefs have been great beneficiaries, as have very rich and moderately rich land and house owners. Investors in transportation have had a field day, as have financial institutions which have cashed in, with some offering credit to further propel accumulation by 'annihilating space through time', to borrow from Harvey's interpretation of Marxian political economy. Some institutions of the state have benefitted, such as the Tourism Authority, which licenses the hospitality sector, but others such as the Ghana Revenue Authority will have to wait to see good oil days, if they will ever come. Even more disturbing, some social groups have been made worse off on the sea and on the land. Fishers have struggled with harvest, not only due to pollution but also as a consequence of restrictive water property rights that have dispossessed them of following fish to certain areas which draw fish away from fishers, further culminating in a heavy cost of operation as more labour power and money have to be expended in obtaining fish even at subsistence levels. Apart from the fishers who are clearly witnessing troubled waters, sea animals have experienced some dislocation and destabilisation, although the evidence here is inconclusive. On the land, farmers have missed out on their lots, some of which have been lost through dispossession. Both fishers and farmers, and indeed to a large degree, poor people generally have suffered evictions by landlords seeking to re-let their houses to people in a higher class, monied, and wealthy people connected to the oil companies.

The experience of evictions is part of the dynamics of a new type of housing that is emerging, if not emergent. New renters are higher income earners. Owners are moving into gated communities. Low-income people are being moved into poorer and poorer housing or paying the same rent for poorer housing or sometimes even more rent for the same housing. Housing-related conflict is on the rise and the type of housing conflict has changed substantially. Once typified by petty quarrels, housing disagreements now centre on issues of recovery of possession for higher class tenants. Estate agents are facilitating this process, but not only the traditional agents but also a new class of agents differentiated by their more economically comfortable and multiracial character. A new class of hotels is also springing up and others are being upgraded. These dynamics are attracting sex workers into areas where entertainment is booming. While media accounts suggest widespread deleterious effects of this sex economy on society, the evidence does not support such accounts. Whether property values would have been higher without the presence of the sex workers is counterfactual and difficult to prove. Yet, hate discourses surround sex workers who can be said to require support, if it is argued that their bodies are being commodified and exploited. These trends are indicative of a general marginalisation of women in the boom experienced in the city, also partly reflected in the types of jobs created and the share that goes to women. Further, the boom has, relative to migrants from other cities and expatriates from other countries, bypassed the local people and offered men further and better opportunities than women. Also disturbing is the further deepening of differences and differentiation between Sekondi-Takoradi and the bigger cities, as jobs go to Accra and Kumasi-based people.

More broadly, the increase in rent has important ramifications. More rent is now paid, quantitatively and arguably proportionally. Sprawl can result as poorer people move to settle at the periphery of the city, leaving the city centre and prime areas for the propertied class. With more rent being demanded and paid, speculation will continue and these dynamics – more rent, more speculation, more rent – can discourage work or extra work in the long run as people get discouraged from working more and more only to spend their wages on rent. More so, it is likely to drive both cost-push and demand-pull inflation as businesses increase the prices of goods and services to reflect growing rent. Or, as people demand more and more of goods that are likely to fall short – a situation that is particularly likely if producers with declining wages (or wages swallowed by rent) get discouraged to produce more. For the national economy as a whole, this process may cause an economic depression to set in, at least from a Georgist perspective. A local, Sekondi-Takoradi specific recession is possible in such circumstances and their repercussions can reverberate in the entire urban economy of Ghana. The key point here is that the existence of rent, arising from the monopolisation of land and natural resources, causes 'progress and poverty' because the common wealth is captured by a landlord class that does not necessarily do anything.

While this effort to analytically unearth and establish these problems and demonstrate how they are experienced by different classes is imperative, it is not enough. For this reason, this book has also discussed practical suggestions that can contribute to enhancing the quality of urban and regional development, and even national development. This aspect of the book is not offered as a panacea or even a prescription. Neither is it assumed that policy makers will simply implement it. Rather, this aspect of the book tries to change the orientation of the current debate that has centred on the macro issues to the neglect of micro ones: local economy, local environment, and urban social change, while stimulating further discussion on effective urban policy strategies that are understood within the broader, global oil industry. For these reasons, the attempt to discuss policy issues is always juxtaposed with a discussion of the structural and institutional impediments to change, not only locally but also internationally.

For example, currently, the city authorities are mandated to ensure harmonious urban economic development. While they seem to be interested in doing so, they face many institutional-structural challenges. Institutionally, they can use the power to elicit betterment payment to bring in more money and to exert a ceiling on conspicuous consumption. Further, there are laws on property taxation that they can invoke in their favour. However, there are institutional setbacks such as lack of personnel or qualified personnel. Further, the tax system is set up in such a way that it does not permit the checking of speculation, as it falls on real estate, not on land per se. So, STMA is not in the position to curtail aggrandisement let alone rake in more revenue from betterment payment. Thus, the resources of the city authorities are slight compared with the challenges ahead. In turn, it has had to rely on the oil companies for its programmes. This decision, while extremely beneficial in an economic sense, falters badly in terms of political economic agendas as the state in this alliance with capital is then forced into a position of compromise when it is dealing with its pay masters. The city is now being planned for the rich and against the poor and minorities by ignoring them or spreading 'hate' discourses about them. Here is the neoliberal city, the rentier city, and the comprador city – to borrow respectively from Harvey, Mahdavy, and Nwoke.

However, these experiences, although naturalised by economists are not natural. Rather, they are socially conditioned, diachronically and synchronically and hence cast doubts over arguments that there is a 'natural resource curse' emasculated from Africa's past and present social context. The situation of weak finances is not a present-day creation alone. Its roots lie in the structural adjustment era where there was a consistent undermining of the state or restructuring it to defend and extend capitalist development. The retrenchment of staff then and a continuing implementation of neoliberal ideology now leading to a new public management bureaucracy and protocols of trying to control public budgets have led to a reluctance of the national state to recruit more skilful and experienced planners. Instead, jobs are subcontracted or privatised. So, we see capitalist and neoliberal forces of change and continuity enveloping the political structures in the city.

The central state supports the local state in some ways such as through allocations from say the District Assemblies Common Fund. However, it also undermines the position of urban authorities. The classic case is its refusal to set aside 10 per cent of oil revenues potentially for the urban authorities to use for development. Rather, the central state claimed that all its people are Ghanaians and have equal entitlement to the resources of the country. On equal and just terms, more generally, the state has little or no enviable record of making good compensation claims made by land-owning communities. While the country's constitution makes it obligatory to pay compensation, it is currently unlikely that the central state will offer compensation to the people who have suffered water and physical land taking.

On a bigger scale, the Ghanaian state has failed to maximise its potential of capturing more oil rents, giving the oil companies the field day to determine what they want to return to the Ghanaian public as rent. The oil companies have succeeded in contracting to pay only a small percentage of oil rents as royalty, and made use of the existing laws to legally avoid tax payments. The fixation on transparency diverts attention from other tensions and contradictions in which the petro industry is embedded. These are more complex than merely trying to find a public administration solution to a political economic imbroglio. From this perspective alone, claims of 'immunity' from political economic tensions related to the oil sector are misleading. Resource curse analyses are fundamentally problematic but also erroneous are attempts to substitute them with 'resource curse immunity' gaze. A transition from the simplistic resource blessing/curse discussion must entail not just a change but a just change. Being man-made, the pressures driven by the oil sector can be unmade.

Any attempt to adopt nationalisation as a policy goal in the short run is likely to be abortive. However, more careful negotiations of state–oil company contracts can help to increase the share of rents the state can extract, as debates expand to looking at not only existing rents but also potential rents and how the state can increase its portion of the various types of rent. Together with the existing rents, three categories of policy have been put forward in this book: the payment of compensation and the extraction of betterment; taxation of land rents, taxation of land values, taxation of capital gains, and taxation of housing rents; and the socialisation of oil rents entailing investment in education, public housing, and public transport – with limited use of direct cash transfers. These options, it has been argued, have the potential to lead to a more effective use of oil rents for public investments that generate employment in various sectors of the economy, provide enhanced human development, and better quality environment, reduce inequality, enhance the quality of life in the metropolis, empower the city authorities to be more autonomous and responsive to the urban residents for whom they work, and ensure the development of other cities and regions in the country.

Taxation is important. It can be used to reduce the rising land and housing prices or even generate revenue for public infrastructural development which, in turn, can create employment. It can help to reduce the dependency of the local government on the centre and on foreign capital. It has great potential to break up land concentration and nip a process of creating *latifundos* in the bud. It is not that taxation proposals were a panacea. Indeed, this work argues that taxation analysis is inadequate without looking at international influences and the role of the state in a historical context, as well as the relationship between or among several arms of the state or, say, between the traditional, local, and national state. Further, the existence of tax laws which are seemingly progressive is necessary but not sufficient to produce progressive ends. Strengthening the capacity of the state is crucially important, as is overhauling the numerous exemptions that characterise seemingly progressive laws.

The strengthening of the state is further justified because of the misleading notion that more markets equals better management. The downsides of poor distribution are apparent as are the issues of actual and potential pollution and actual displacement under a predominantly privately owned resource wealth. The evidence shows that more markets create worse distribution of resources. Urban authorities can help to steer harmonious urban economic development. The recruitment of highly alert urban planners and investment in the improvement of their capacity and those of other professionals ancillary to planning can be helpful too. All the effort to put in place (good) governance structures is incomplete without empowering these authorities (including the rent control department and the Ghana Revenue Authority) through support in logistics and personnel, information building, and technical planning support. Strengthening the local state can make its plans more inclusive too.

State provision of education using the support of oil rents is vastly important, even for its own sake. It is also important for effective public and private development, as a more educated population will be in a better position to collectively and individually take decisions and actions to help themselves and others, the twin city, the region, and the country to better advance its course, better achieve its needs, and better set new goals and plans for the future. For governance too, a critically educated demos can better hold the government to account and offer alternatives to public policy. The educated demos can gain access to employment opportunities, but such policies ought to be regarded as complementary to others. Connected to the recommendation of providing quality education is how empowerment through education can help the local population to stand up for their rights as workers, drawing on the labour laws in the country. So, overall, investing in public education is one effective area for the utilisation of oil rents.

The expansion in public housing is worth considering too. The housing–social–health–economic development nexus is well established. Inputs in the housing sector, the process of housing construction, and the services needed by the sector all generate jobs which can be harnessed for struggling local people who are yet to benefit from either direct or indirect employment generated by the oil sector. More directly, building more housing will help the great inequality that private housing is currently generating, curb the increasing trend of evictions, and help reduce the housing deficit not only in the city but also nationally. These advantages can be further enhanced if the public housing programme is linked with local industries that will produce the construction, electrical, masonry, plumbing, and carpentry materials needed to complete the housing programme.

Investment in public transport will help to generate employment, and stimulate the development of the wood and steel industries from which some of the inputs to develop the rail sector can be obtained. Upholstery industry will boom, as the need for seat covers in the rail sector increases and with these increases will be employment. Further, when developed, the rail sector will enhance urban mobility within and outside the Sekondi-Takoradi Metropolis. Also, its effects on the environment are vastly more propitious, as concern about investment ought not to ignore the environment, and social. Indeed, environmental, political, and economic issues should not be put in silos. It is crucially important to begin to think about merging the Ministry of Environment, Science, Technology, and Innovation, the Ministry of Energy and Petroleum, and the Ministry of Finance and Economic Planning to realise the interconnections between the economy, society, and environment. Still for rail, the activism of its union ought to be seen positively, as this can help to check excesses in the oil-economy-environment and urban change connections. Indeed, the activities of the union can be applied by labour elsewhere and to sectors of the urban society and economy, such as fishers.

These policy implications have not been 'proved' as perfect or without blemish. What they seek to do is to widen current discourses rather than limit or delimit them to a few orthodox social scientific perspectives. Even then, these policy options face the practical concern of how they are to be adopted. The shift or broadening up is not going to be easy, as the overwhelming discourse is on national governance. Progress has been made, though, with the adoption of a National Urban Policy, even if the policy does not sufficiently anticipate oil. The few studies cited along the way also give some hope that research on urban governance and oil are likely to flourish. Similarly, the rise of the class of planning consultants working with local planners gives some hope. However, there is need for greater, genuine collaboration between the local interests, the national sectors, and the international forces. Working at the local scale alone, without engaging the national and international, can lead to tokenism in public policy making. Looking at the national and international alone around the notion of an ahistoric, aspatial framework called 'good governance' grossly misunderstands the problem. If this holistic view, sensitive to the local concerns were taken, it would follow that a substantial percentage of the resource revenues could be set aside for the development of the oil city.

Oil is a dynamic substance, flowing at local and through national, regional, and sub-regional spaces but traded internationally. So, oiling the urban economy, questions of capital, land, labour, and the state are to be emphasised at the local level but appreciated at the national, regional, sub-regional, and international scales, to understand their embeddedness within circuits of capital and institutions of power even at the local traditional level. These levels and levers of intervention clearly mean that urban political economic analysis can best forcefully throw light on local and urban dynamics if it does not overlook the regional, national, sub-regional (that is, West African), and international processes of capital accumulation. The evidence analysed in this book, showing considerable movement between cities and regions, and neighbouring regions demonstrates that the resource curse concept and its resulting depoliticised policy orientation is highly inadequate.

However, the impediments to change will be numerous and, among others, contingent on the nature of the political class, the actions of civil society groups, and the media, the type of scholarly studies on the topic and how these are disseminated. In short, the politics of change will be contentious and complex. The dynamic relationship between land, labour, and capital will require the installation of a particular state that is oriented towards progressive social change and is prepared to undo the vestiges of colonialism while taking a bold stance against the free market ideology that is the usual part of the policies of international development agencies. Of course, capital is adept at aligning itself with the state to obtain the best conditions of production, so such a bold stance will not escape attempts at being compromised.

The book has shown that the Ghanaian state does not consistently pursue the public interest, neither does it always pursue its own interests. Aspects of both tendencies can be seen in its role in the oil era, but it has a propensity to systematically enforce particular class interests. It is, therefore, a comprador state, a rentier state, and a neoliberal state, drawing on the ideas of Harvey, Mahdavy, and Nwoke. Yet, the massive corruption that normally accompanies such tendencies is minimal in the Ghanaian case. Indeed, it possesses an internationally acknowledged positive record of upholding human rights and providing room for dissent. Also, the Ghanaian state is known to be growth centric in its economic management programme. So, one might be tempted to conclude that the Ghanaian example is one of a rentier state without rentier characteristics. Yet, so far, there has been a particular post-colonial and neoliberal structure masquerading as a 'social democratic' state which is in charge of the management of oil rents and, as we have seen, is more favourable to capital, the landed, and propertied classes, than the marginalised, minorities, and the poor. Its record at pursuing sustainable local environment has been anything but stellar.

The political alternatives currently on offer are mainly pretentious 'social democratic' groups and a party professing 'property owning democracy' with sympathies of the neoliberal order. The party that was swept to power on the eve of independence has consistently provided much better progressive alternatives in the public sphere, but it is bogged and weighed down by internal disunity and petty squabbles such that it is organisationally weak to win power. Not all is lost, however, as civil society, the media, and academics continue to show considerable interest in oil politics and the political economy of oil. Nevertheless, the prevailing understanding of the issues – 'oil blessing', 'oil curse', 'anomie', 'rentier state', or 'oil as enclave activity' – is partial and unhelpful. Further, the neoliberal character of the existing civil society groups, even if sometimes unintended, is a blot on their ability to drive not just change; but a just change. Hopefully, the analyses in *Oiling the Urban Economy: Land, labour, capital, and the state in Sekondi-Takoradi, Ghana*, looking at the thesis and antithesis and providing original insights and synthesis, will contribute to shedding more light on the issues that are central to the processes of urban, regional, and national economic development under the influence and expectations of oil exploration, production, and exportation.

Appendix

Table A.1 Record of interviews and data gathered

Institution	Number interviewed	Description of respondents	Month of interview	Nature of data and how they were gathered.
Ghana Tourism Authority (Western Region)	3	Acting Regional Manager and 2 of his assistants	December 2012	Data on investment trends in commercial urban real estate. These data were taken from the files of the authority and supplemented by detailed oral explanations through detailed interviews.
Takoradi Polytechnic	2	Acting Dean, and Vice Dean International Programmes	December 2012 to January 2013	Data on the number and profile of students enrolling in oil and gas courses.
Oil and Gas Training and Recruitment Centre	2	Business Development Officer and the Officer-in-Charge of training	March 2013	Data on the number and profile/origin of students enrolling in oil and gas courses.
Ghana Revenue Authority (Takoradi)	3	Chief Revenue Officer and 2 Revenue Officers	December 2012	Data on rent taxation. These data were pieced together from several files and loose sheets. The process of compilation was supervised and assisted by the Chief Revenue Officer and number of assistants – all of whom provided further oral explanation when needed.
Ghana Highways Authority and National Road Safety Commission (Sekondi-Takoradi)	2	Assistant Planning Officer (National Road Safety Commission); and 1 Senior Officer at the Ghana Highways Authority	January 2013	Data on urban transportation. These data were taken from various reports held and provided by these agencies with helpful explanations when asked.

continued

Table A.1 Continued

Institution	Number interviewed	Description of respondents	Month of interview	Nature of data and how they were gathered.
Lands Commission, Lands and Land Valuation Division (Sekondi-Takoradi)	5	Head of Valuation Division, Head of Mapping Division, 2 Senior Land Surveyors, and 1 Land Economist	January 2013	Data on land values and other trends in the land market. These data were of two kinds. The first came through interviews to record the experienced and considered reflections of the surveyors and land economists who have been working as such in the metropolis. The second was a mixture of oral and documentary evidence.
Rent Control Department	4	Acting Chief Rent Officer, with oversight responsibility of the Rent Control Department in Ghana, Senior Rent Officer, heading the Sekondi-Takoradi department, and 2 officers/assistants chosen because they are directly in charge of compiling records on a number of rent-related issues, such as rental payments and conflicts	December 2012 to January 2013	Data on the rental market dynamics, conflict, and other issues arising from the sector. Detailed interviews with all 3 officers who provided oral and documentary evidence on the question about trends in the rental market sector. The data were gathered from the files of the department with the help of especially 2 junior officers.
Ahantaman Rural Bank	3	Head of Credit and Credit Officer (Agona), and Manager (Takoradi)	January 2013	Data on credit for commercial urban transportation. The data were gathered through oral interviews backed sometimes by documentary evidence.
Driver and Vehicle Licensing Authority	1	Senior Executive Officer	January 2013	Data on urban transport. Data here were mainly from official files and supplemented briefly by the officer when it was clear the data in the files did not show clear trends of transport population.

Institution	No.	Respondents	Date	Data
Sekondi-Takoradi Metropolitan Assembly	4	Head of Physical Planning Division, 1 retired Physical Planning Officer, and 2 Planners/Planning Officers	December 2012 to March 2013	Data on various planning issues in Sekondi-Takoradi obtained orally at first but later supplemented with reports and other planning documents.
Osiakwan Baah Ltd and Ussell Services	2	Estate agents	January 2013	Data on the rental market. These data were mainly oral and supplemented only by the documentation of prices/rates/rents recorded on signboards and, in the case of Ussell, brochures. The extensive interviews, however, shed light on the housing and rental market in the metropolis.
Asenta Properties Ltd	1	Sekondi-Takoradi Head of Asenta Properties Ltd, the second oldest valuation and real estate firm in Takoradi	December 2012 to March 2013	Data on the property market generally. An experienced valuer, a member of the Ghana Institution of Surveyors, and one of the very few valuers in the metropolis, this interviewee discussed data from previous respondents, and provided valuable insights on trends in property values in the metropolis and drew my attention to other markers of property prices such as those recorded by agents on signboards nailed to trees and scattered around the metropolis.
Takoradi Oil Village	4	Property developer (1), Architects (2), and Project Officer (1)	February 2013	Data on the target market for this gated housing community, price range, features of the estate, and how it was developed.
Chieftaincy institution	1	Paramount Chief of Esikado, Sekondi	March 2013	Data on land, chieftaincy, and history of Sekondi-Takoradi.

continued

Table A.1 Continued

Institution	Number interviewed	Description of respondents	Month of interview	Nature of data and how they were gathered.
Ghana Railway Company	4	All long serving members of staff of the company made of 2 former union leaders	January to March, 2013	History of the railway company, workings of the union, and present challenges of the company.
Ghana Police Service (Sekondi-Takoradi)	2	Regional Commander and 1 Constable	September 2013	Data on crime in Sekondi-Takoradi, especially Vienna City area [I thank my research assistant for helping to conduct these interviews].

Bibliography

Aaron K.K., 2005, 'Big oil, rural poverty, and environmental degradation in the Niger Delta region of Nigeria', *Journal of Agricultural Safety and Health*, vol. 11, no. 2, pp. 127–134.

Abdulai T.R., 2010, *Traditional Landholding Institutions in Sub-Saharan Africa, the Operation of Traditional Landholding Institutions in Sub-Sahara Africa: A Case Study of Ghana*, Lambert Academic Publishing, Saarbrücken.

Ablakwa S., 2010, 'Statement read by the deputy minister for information on the management of oil revenue', Accra, 2 December.

ABLIN Consult, 2008, 'A study on the conditions of service of commercial vehicle drivers and impact on road safety conditions', ABLIN Consult, Accra.

Ablo A.D., 2012, 'Manning the rigs: A study of offshore employment in Ghana's oil industry', MPhil thesis submitted to the Department of Geography, University of Bergen, Norway.

Aboagye G., 2011, 'China wants more oil from Ghana', *Offline*, March, pp. 11–12.

Achiaw N., 2008, 'Oil boom alert … prez charges stakeholders to make it a blessing not a curse (1a)', Graphic Front Page Stories, 26 February. Available at: http://tettehamoako.blogspot.com/2008/02/oil-boom-alert-prez-charges.html (accessed 29 October 2008).

Ackah-Baidoo A., 2012, 'Enclave development and "offshore corporate social responsibility": Implications for oil-rich sub-Saharan Africa', *Resources Policy*, vol. 37, no. 2, pp. 152–159.

Ackah-Baidoo A., 2013, 'Fishing in troubled waters: Oil production, seaweed and community level grievances in the Western Region of Ghana', *Community Development Journal*, vol. 48, no. 3, pp. 406–420.

Acosta A. and Heuty A., 2009, 'Can Ghana avoid the oil curse? A prospective look into natural resource governance', Policy Briefing prepared for the UK's Department for International Development (DfID).

Addo J.O., 2013, 'Regeneration with conservation and preservation or bust', *ArchiAfrika Magazine*, vol. 3, August.

Addo S.T., 2006, *Geography Transport and Development: A Spatial Trinity*, Ghana Universities Press, Accra.

Administrators, 2012, 'Colonial history (Sekondi-Takoradi)'. Available at: www.ghanaoilcity.com (accessed 11 July 2012).

Adu G., 2009, 'On the theory of optimal depletion of an exhaustible resource: The case of oil in Ghana', *African Review of Economics and Finance*, vol. 1, no. 1, pp. 40–50.

African Development Bank and African Union, 2009, *Oil and Gas in Africa*, Oxford University Press, New York.

African Development Bank (AfDB), OECD Development Centre, Economic Commission for Africa (ECA), and the UN Development Programme (UNDP), 2013, 'African economic outlook', OECD Development Centre, Issy les Moulineaux, France.

Africa Progress Panel, 2013, *African Progress Report 2013*, Africa Progress Secretariat, Geneva.

Afrikan Post, 2009, 'History of oil discovery in Ghana: The EO groups role'. Available at: http://news.myjoyonline.com/features/200911/38360.asp (accessed 22 August 2010).

Agbefu R.M.E., 2011, 'Discovery of oil in Ghana: Meeting the expectations of local people', MPhil thesis in Culture, Environment and Society, Centre for Development and the Environment, University of Oslo, Blindern, Oslo.

Agovi K., 2008, 'A king is not above insult: The politics of good governance in Nzema avudwene festival songs', in Furniss G. and Gunner L., eds, *Power, Marginality and African Oral Literature*, Cambridge University Press, Cambridge, pp. 47–61.

Agular J., 2012, '26 communities benefit from Jubilee LEED', *Daily Graphic*, 30 November.

Aigbokhan B., 2008, 'Growth, inequality and poverty in Nigeria', Economic Commission for Africa, ACGS/MPAMS Discussion Paper No. 3.

Akabzaa T. and Darimani A., 2001, 'Impact of mining sector investment in Ghana: A study of the Tarkwa mining region', Draft report prepared for *SAPRI*. Available at: www.saprin.org/ghana/research/gha_mining.pdf (accessed 9 June 2009).

Akhaine S.O., 2010, 'Nigeria: Politics and the end of oil', *Review of African Political Economy*, vol. 37, no. 123, pp. 89–91.

Akli E., 2010, 'Ghana oil sold for $67', *The Chronicle*, 21 January.

Aklorbortu M.D., 2013, 'Ghana's marine environment under threat as more whales are washed ashore', *Daily Graphic*, 5 September.

Akufo-Addo N., 2010, 'The economic prospects of the oil discovery to the nation', Speech delivered at the 39th Faculty Week Celebration of the Faculty of Social Sciences, KNUST, Kumasi, 6 October.

Akufo-Addo N., 2012, 'Education: The key to transforming Ghana', Speech delivered by the NPP 2012 Presidential candidate, Nana Akufo-Addo, on 26 November 2012, University of Cape Coast, Cape Coast.

Akyeampong E., 1994, 'The state and alcohol revenues: Promoting "economic development" in Gold Coast/Ghana, 1919 to the present', *Social History*, vol. 27, no. 4, pp. 393–411.

Akyeampong E., 1997, 'Sexuality and prostitution among the Akan of the Gold Coast *c.*1650–1950', *Past and Present*, vol. 156, pp. 144–173.

Akyeampong K.E., 2003, 'Review of soldiers, airmen, spies and whisperers: The Gold Coast in World War II', *Journal of Military History*, vol. 67, no. 3, pp. 971–972.

Al-Mubarak F., 1999, 'Oil, urban development and planning in the eastern province of Saudi Arabia: The case of the Arab American Oil Company in the 1930's–1970's, *J. King Saud Univ.*, Vol. 11, *Arch. & Planning*, pp. 31–51 (A.H. 1419/1999).

Al-mulali U., 2010, 'The impact of oil prices on the exchange rate and economic growth in Norway', MPRA Paper, no. 26257.

Al-Nakib F., 2013, 'Kuwait's modern spectacle: Oil wealth and the making of a new capital city, 1950–90', *Comparative Studies of South Asia, Africa and the Middle East*, vol. 33, no. 1, pp. 7–25.

Albrecht S.L., 1982, 'Commentary on "local social disruption and western energy development: A critical review"', *Pacific Sociological Review*, vol. 25, no. 3, pp. 297–306.

Alchian A.A. and Demsetz H., 1973, 'The property right paradigm', *Journal of Economic History*, vol. 33, no. 1, pp. 16–27.

Ali O.K., Hashim M., Rostam K., and Jusoh H., 2008, 'Changes in residential land-use of Tripoli city, Libya: 1969–2005', *Malaysian Journal of Society and Space*, vol. 4, pp. 71–84.

Alissa R., 2013, 'The oil town of Ahmadi since 1946: From colonial town to nostalgic city', *Comparative Studies of South Asia, Africa and the Middle East*, vol. 33, no. 1, pp. 41–58.

Allan J., 2009, 'Can Ghana survive the oil boom or doom? *Ghanaian Chronicle*, 24 December.

Allum P., 2010, 'Oil offers hope of middle-income status for Ghana', *IMF Survey Magazine*, 17 February.

Alterman R., 2011, 'Takings', Paper presented at the Land Resource Compensation Symposium, UTS, Asia-Pacific Centre for Complex Real Property Rights, Sydney, 11–12 July.

Alterman R., 2012, 'Land use regulations and property values: The "windfalls capture" idea revisited', in Brooks N., Donaghy K., and Knaap G.-J., eds, *The Oxford Handbook of Urban Economics and Planning*, Oxford University Press, Oxford, pp. 755–786.

Amin S., 1972, 'Underdevelopment and dependence in Black Africa: Historical origin', *Journal of Peace Research*, vol. 9, pp. 105–119.

Amin S., 2002, 'Africa: Living on the fringe', *Monthly Review*, March, pp. 41–50.

Amoako-Tuffour J. and Ghanney M.A., 2013, 'Ghana's petroleum revenue management law: A social contract for good economic governance and possible challenges', in Appiah-Adu K., ed., *Governance of the Petroleum Sector in an Emerging Developing Economy*, Gower Applied Research, Ashgate, Surrey and Burlington VT, pp. 35–63.

Amoako-Tuffour J. and Owusu-Ayim J., 2010, 'An evaluation of Ghana's petroleum fiscal regime', *Ghana Policy Journal*, vol. 4, December, pp. 7–34.

Amoakohene, M.I., 2006, 'Political communication in an emerging democracy: A comparative analysis of media coverage of two presidential administrations in the fourth republic of Ghana', PhD thesis, University of Leicester, Leicester.

Amoasah G., 2010, 'The potential impacts of oil and gas exploration and production on the coastal zone of Ghana: An ecosystem services approach', MSc thesis in Environmental Sciences, Wageningen University, Wageningen, the Netherlands.

Ampene E., 1965, 'A study in urbanization: Progress report on Obuasi project. A profile on music and movement in the Volta region Part I', *Journal of the Institute of African Studies: Research Review*, vol. 3, nos. 1/2, pp. 42–47.

Anarfi K.J., 1997, 'Vulnerability to sexually transmitted disease: Street children in Accra', *Health Transition Review*, Supplement to vol. 7, pp. 281–307.

Andrews N., 2013, 'Community expectations from Ghana's new oil find: Conceptualizing corporate social responsibility as a grassroots-oriented process', *Africa Today*, vol. 60, no. 1, pp. 55–75.

Anghie A., 1993, ' "The heart of my home": Colonialism, environmental damage, and the Nauru case', *Harvard International Law Journal*, vol. 34, no. 2, pp. 445–506.

Aning K.E., 2011, 'Security and crime management in the wake of the discovery and exploration of oil & gas in Ghana', Avoiding the oil curse in Ghana, International conference held between 24 and 27 November, Koforidua.

Aning S., 2013, 'Oil and gas security issues', in Appiah-Adu K., ed., *Governance of the Petroleum Sector in an Emerging Developing Economy*, Gower Applied Research, Ashgate, Surrey and Burlington VT, pp. 233–247.

Appiah-Adu K., 2013, 'Conclusion', in Appiah-Adu K., ed., *Governance of the Petroleum Sector in an Emerging Developing Economy*, Gower Applied Research, Ashgate, Surrey and Burlington VT, pp. 285–292.

Appiah-Adu K. and Sasraku F.M., 2013, 'Revenue management in the oil and gas sector', in Appiah-Adu K., ed., *Governance of the Petroleum Sector in an Emerging Developing Economy*, Gower Applied Research, Ashgate, Surrey and Burlington VT, pp. 27–34.

Appiahene-Gyamfi J., 2003, 'Urban crime trends and patterns in Ghana: The case of Accra', *Journal of Criminal Justice*, vol. 31, pp. 13–23.

Apraku K.K., 2000, 'Contribution to the parliamentary debate on the Internal Revenue Bill', *Parliamentary Debates: Official Report*, 4 October.

Armah-Attoh D. and Awal M., 2012, 'Tax administration in Ghana: Perceived institutional challenges', Draft Afrobarometer Briefing Paper No. XXX, 6 August.

Arku, G. and Harris, R., 2005, 'Housing as a tool of economic development since 1929', *International Journal of Urban and Regional Research*, vol. 29, no. 4, pp. 895–915.

Arowesegbe J., 2009, 'Violence and national development in Nigeria: The political economy of youth restiveness in the Niger Delta', *Review of African Political Economy*, vol. 36, no. 122, pp. 575–594.

Asafu-Adjaye J., 2012, 'Tracking transparency and accountability in Ghana's oil and gas industry', *2011 PTRAC Report*, Institute of Economic Affairs, Accra.

Asamoah J., 2013, 'Ghana's oil city comes alive', *Daily Graphic*, 19 August. Available at: http://graphic.com.gh/features/ghanas-oil-city-comes-alive.html (accessed 14 October 2013).

Asamoah-Adu C., Khonde N., Avorkliah M., Bekoe V., Alary M., Mondor M., Frost E., Deceunink G., Asamoah-Adu A., and Pepin J., 2001, 'HIV infection among sex workers in Accra: Need to target new recruits into the trade', *JAIDS*, vol. 28, pp. 358–366.

Asante S.K.B., 1960, 'The neglected aspects of the activities of the Gold Coast aborigines rights protection society', *Phylon*, vol. 36, no. 1, pp. 32–45.

Asante S.K.B., 1975, 'The neglected aspects of the activities of the Gold Coast aborigines rights protection society', *Phylon*, vol. 36, no. 1, pp. 32–45.

Ashun A., 2004, *Elmina, the Castle and the Slave Trade*, Nyakod Printing Works, Cape Coast.

Asiamah H., 2012, 'Understanding climate change: Delayed construction of Ghana's oil and gas plant has heat implications', *Public Agenda*, 30 January.

Asmah E., 2008, 'Assessing the links between energy services and the MDGs: Is a MAMS application for Ghana possible?' Interim Paper presented at the CSAE Conference 2008 on 'Economic Development in Africa' at St Catherine's College, Oxford, 16–18 March.

Atalay S., 2008, 'Multivocality and native archaeologies', in Habu J., Fawcett C., and Matsunaga J.M., eds, *Evaluating Multiple Narratives: Beyond Nationalist, Colonialist, Imperialist Archaeologies*, Springer, New York, pp. 29–44.

Atash F., 2007, 'The deterioration of urban environments in developing countries: Mitigating the air pollution crisis in Tehran, Iran', *Cities*, vol. 24, no. 6, pp. 399–409.

Athelstan A. and Deller R., 2013, 'Editorial: Visual methodologies', *Graduate Journal of Social Science*, vol. 10, no. 2, pp. 9–19.

Atkinson A., 2007a, 'Cities after oil – 1: "Sustainable development" and energy futures', *City*, vol. 11, no. 2, pp. 201–213.

Atkinson A., 2007b, 'Cities after oil – 2: Background to the collapse of "modern" civilisation', *City*, vol. 11, no. 3, pp. 293–312.

Atkinson A., 2008, 'Cities after oil – 3: Collapse and the fate of cities', *City*, vol. 11, no. 2, pp. 201–213.

Atkinson A., 2009, 'Cities after oil (one more time)', *City*, vol. 13, no. 4, pp. 493–498.

Atkinson A., 2012, 'Urban social reconstruction after oil', *International Journal of Urban Sustainable Development*, vol. 4, no. 1, pp. 94–110.

Atuguba R., 2006, 'The tax culture of Ghana', Research report prepared for the Revenue Mobilisation Support (RMS), Draft Final Report, February.

Austin G., 2005, *Labour, Land and Capital in Ghana: From Slavery to Free Labour in Asante, 1807–1956*, University of Rochester Press, New York.

Auty R., 1993, *Sustaining Development in Mineral Economies: The Resource Curse Thesis*, Routledge, London.

Auty R., 2001, 'The political state and the management of mineral rents in capital-surplus economies: Botswana and Saudi Arabia', *Resources Policy*, vol. 27, pp. 77–86.

Auty R., 2007, 'Natural resources, capital accumulation and the resource curse', *Ecological Economics*, vol. 61, pp. 627–634.

Auty R. and Pontara N., 2008, 'A dual-track strategy for managing Mauritania's projected oil rent', *Development Policy Review*, vol. 26, no. 1, pp. 59–77.

Awuah K.G.B., Hammond F.N., Lamond J.E., and Booth C., 2014, 'Benefits of urban land use planning in Ghana', *Geoforum*, pp. 37–46.

Ayelazuno J., 2011, 'Continuous primitive accumulation in Ghana: The real-life stories of dispossessed peasants in three mining communities', *Review of African Political Economy*, vol. 38, no. 130, pp. 537–550.

Ayelazuno, J., 2013, 'Oil wealth and the well-being of the subaltern classes in Sub-Saharan Africa: A critical analysis of the resource curse in Ghana', *Resource Policy*, http://dx.doi.org/10.1016/j.resourpol.2013.06.009.

Ayitey J.Z., Kidido J.K., Tudzi E.P., 2011, 'Compensation for land use deprivation in mining communities, the law and practice: Case study of Newmont Gold Ghana Limited', *Ghana Surveyor*, vol. 4, no. 1, pp. 32–40.

Ayittey S., 2010, 'Press briefing on Ghana's environmental preparedness for first oil', Statement made in the Ministry of Information Conference room, Accra, 2 September.

Bacon R. and Kojima M., 2008, *Coping with Oil Price Volatility*, World Bank, Washington DC.

Badgley C., 2011, 'Fishing and the offshore oil industry: A delicate imbalance', Top Stories reported by The Center for Public Integrity, 10 June.

Baidoo J., 2012, 'Veep unveils prototypes of two cities', *Daily Graphic*, 9 July. Available at: www.graphic.com.gh (accessed 14 July 2012).

Banchirigah S.M., 2008, 'Challenges with eradicating illegal mining in Ghana: A perspective from the grassroots', *Resources Policy*, vol. 33, no. 1, pp. 29–38.

Baran P., 1957, *The Political Economy of Growth*, Monthly Review Press, New York.

Barbier B.E., 2005, *Natural Resources and Economic Development*, Cambridge University Press, New York.

Basedau M. and Mehler A., eds, 2005, *Resource Politics in Sub-Saharan Africa*, Institute of African Studies, Hamburg.

Bawole J.N., 2013, 'Public hearing or "hearing public"? An evaluation of the participation of local stakeholders in environmental impact of Ghana's Jubilee oil fields', *Environmental Management*, vol. 52, no. 2, pp. 385–397.

Beblawi, H., 1987, 'The rentier state in the Arab world', in Beblawi, H. and Luciani, G. eds, *The Rentier State*, North Ryde, NSW, Istituto Affari Internazionali, pp. 49–62.

Beine M., Bos C.S., Coulombe S., 2012, 'Does the Canadian economy suffer from Dutch disease?' *Resource and Energy Economics*, vol. 34, no. 4, pp. 468–492.

Bennion F., 1962, *Constitutional Law of Ghana*, Butterworths, London.

Bet-Shlimon A., 2013, 'The politics and ideology of urban development in Iraq's oil city, Kirkuk, 1946–58', *Comparative Studies of South Asia, Africa and the Middle East*, vol. 33, no. 1, pp. 26–40.

Bimpong-Buta S.Y., 2005, 'The role of the supreme court in the development of constitutional law in Ghana', LLD thesis submitted to the University of South Africa, South Africa.

Bindman J. and Doezeman J., 1997, 'Redefining prostitution as sex work on the international agenda', Network of Sex Work Projects. Available at: www.walnet.org/csis/papers/redefining.html (accessed 19 January 2013).

Bloch A. and Owusu-G., 2011, 'Linkages in the gold mining industry: Challenging the enclave thesis', Making the Most of Commodities Programme (MMCI) Discussion Paper 1, Open University, Cape Town, March.

Bloch A. and Owusu-G., 2012, 'Linkages in the gold mining industry: Challenging the enclave thesis', *Resources Policy*, vol. 37, pp. 434–442.

Boahen A., 1964, 'Review of the roots of Ghanaian nationalism: A political history of Ghana, 1850–1928 by David Kimble', *Journal of African History*, vol. 5, no. 1, pp. 127–132.

Boahen A., 2000, *Ghana: Evolution and Change in the Nineteenth and Twentieth Centuries*, Sankofa Educational Publishers Ltd, Accra.

Bob-Milliar G., 2009, 'Chieftaincy, diaspora, and development: The institution of *Nksuohene* in Ghana', *African Affairs*, vol. 108, no. 433, pp. 541–558.

Bob-Milliar G., 2011, 'Political party activism in Ghana: Factors influencing the decision of the politically active to join a political party', *Democratization*, vol. 19, no. 4, pp. 668–689.

Boianovsky M., 2011, 'Humboldt and the economists on natural resources, institutions and underdevelopment (1752 to 1859)', *European Journal of the History of Economic Thought*, vol. 20, no. 1, pp. 58–88.

Boohene R. and Peprah J.A., 2011, 'Women, livelihood and oil and gas discovery in Ghana: An exploratory study of Cape Three Points and surrounding communities', *Journal of Sustainable Development*, vol. 4, no. 3, pp. 185–195.

Boudet H.S. and Ortolano L., 2010, 'A tale of two sittings: Contentious politics in liquefied natural gas facility sitting in California', *Journal of Planning Education and Research*, vol. 30, no. 5, pp. 5–21.

Bougrine H., 2006, 'Oil: Profits of the chain keepers', *International Journal of Political Economy*, vol. 35, no. 2, pp. 35–53.

Boydell S. and Baya U., 2011, 'Equitable compensation models', Paper presented at the Land Resource Compensation Symposium, UTS, Asia-Pacific Centre for Complex Real Property Rights, Sydney, 11–12 July.

Boye C., 2010, 'The oil flows', *Ghanaian Times*, 16 December.

Boyefio G., 2012, 'Oil boost night life in Takoradi'. Available at: http://gilly2002gh.blogspot.com.au/2012/08/oil-boost-night-life-in-takoradi.html (accessed 17 March 2012).

Boyer H., Beatley T., and Newman P., 2009, *Resilient Cities Responding to Peak Oil and Climate Change*, Island Press, Washington DC.

Bratton M. and van de Walle N., 1994, 'Neopatrimonial regimes and political transitions in Africa', *World Politics*, vol. 46, no. 4, pp. 453–489.

Breisinger C., Diao X., Schweickert R., and Wiebeltt M., 2010, 'Managing future oil revenues in Ghana: An assessment of alternative allocation options', *African Development Review*, vol. 22, no. 2, pp. 303–315.

Brobbey K., 1990, 'Customary tenure and locus standi in compensation claims in Ghana' *Journal of Property Valuation and Investment*, no. 9, pp. 313–322.

Bromley D., 'Constitutional political economy: Property claims in a dynamic world', *Contemporary Economic Policy*, vol. xv, pp. 43–54.

Bruno S.F., 2008, 'City, culture and happiness', *Ethos*, June, pp. 102–111.

Bryceson, D., 2011, 'Birth of a market town in Tanzania: Towards narrative studies of urban Africa', *Journal of Eastern African Studies*, vol. 5, no. 2, pp. 274–293.

Bryceson D. and MacKinnon D., 2012, 'Eureka and beyond: Mining's impact on African urbanisation', *Journal of Contemporary African Studies*, vol. 30, no. 4, pp. 513–537.

Bryceson D., Jonsson J.B., and Sherrington R., 2010, 'Miners' magic: Artisanal mining, the albino fetish and murder in Tanzania', *Journal of Modern African Studies*, vol. 48, no. 3, pp. 353–382.

Bryceson D.F. and Jønsson J.B., 2012, 'Tanzanian artisanal gold mining: Present and future', British-Tanzania Society (BTS) Presentation, SOAS, London, 15 March.

Building and Road Research Institute (BRRI), 2012, *Road Traffic Crashes in Ghana: Statistics 2011 (Final)*, National Road Safety Commission, Accra.

Busia K.A., 1950, *A Report on a Social Survey of Sekondi-Takoradi*, Crown Agents for the Colonies for the Govt of the Gold Coast, London.

Butler G., Jones E., and Stilwell F., 2009, *Political Economy Now! The Struggle for Alternative Economics at the University of Sydney*, Darlington Press, Sydney.

Carmignani F. and Chowdhury A., 2011, 'The development effects of natural resources: A geographical dimension', William Davidson Institute Working Paper No. 1022.

Castle E.N., 1965, 'The market mechanism, externalities, and land economics', *Journal of Farm Economics*, vol. 47, no. 3, pp. 542–556.

Cavnar A., 2008, 'Averting the resource curse in Ghana: The need for accountability', *CDD-Ghana Briefing Paper*, vol. 9, no. 3, pp. 1–4.

Centre for Policy Analysis (CEPA), 2010, 'The economy of Ghana', Mid-term report launch statement, CEPA, Accra.

Centre for Policy Analysis (CEPA), 2011, 'Managing an oil economy', *CEPA News*, 25 January.

Chalfin B, 2008, 'Cars, the customs service, and sumptuary rule in neoliberal Ghana', *Comparative Studies in Society and History*, vol. 50, no. 2, pp. 424–453.

CHF International, 2010, *Sekondi-Takoradi Poverty Map: A Guide to Urban Poverty Reduction in Sekondi-Takoradi*, CHF International, Accra.

Chronicle, 2013, 'Delayed Ghana gas and high cost of crude power generation', *Chronicle*, 1 July. Available at: http://thechronicle.com.gh/?p=58091 (accessed 10 October 2013).

Coastal Resources Center, 2010, *Hen Mpoano: Our Coast, Our Future, Western Region of Ghana*, Coastal Resources Center, University of Rhode Island, Narragansett RI.

Colgan J., 2011, 'Oil and resource-backed aggression', *Energy Policy*, vol. 39, no. 3, pp. 1669–1676.

Collier P., 2003, 'The market for civil war', *Foreign Policy*, May/June, no. 136, pp. 38–45.

Collier P., 2006, 'African growth: Why a "big push"?' *Journal of African Economies*, vol. 15, AERC Supplement 2, pp. 188–211.

Collier P., 2007, 'Managing commodity booms: Lessons of international experience', Paper prepared for the African Economic Research Consortium Centre for the Study of African Economies, Department of Economics, Oxford University, Oxford.

Collier P., 2008, *The Bottom Billion: Why the Poorest Countries are Failing and What Can be Done About it*, New York, Oxford University Press.

Collier P., 2009, *Wars, Guns and Votes: Democracy in Dangerous Places*, Bodley Head, London.

Consortium, 2012, 'Interim report: The preparation of Sekondi-Takoradi structure plan', The Consortium, Accra.

Convention People's Party (CPP), 2012, *Convention People's Party Manifesto 2012*, CPP, Accra.

Cooley A., 2001, 'Booms and busts theorizing institutional formation and change in oil states', *Review of International Political economy*, vol. 8, no. 1, pp. 163–180.

Corden M. and Neary P., 1982, 'Booming sector and de-industrialisation in a small open economy', *Economy Journal*, vol. 92, no. 368, pp. 825–848.

Correspondent, 1943, 'Takoradi: A memory, Scottish', *Geographical Magazine*, vol. 59, no. 1, pp. 37–38.

Crawford G., 2010, 'Decentralisation and struggles for basic rights in Ghana: opportunities and constraints', *International Journal of Human Rights*, vol. 14, no. 1, pp. 92–125.

Crisp J., 1979, 'Review of class power and ideology in Ghana: the railway of Sekondi', *Review of African Political Economy*, vol. 14 (January–April), pp. 111–113.

Crook R., 2003, 'Decentralisation and poverty reduction in Africa: The politics of local-central relations', *Public Administration and Development*, vol. 23, pp. 77–88.

Cui Z., 2011, 'Partial intimations of the coming whole: The Chongqing, experiment in light of the theories of Henry George, James Meade, and Antonio Gramsci', *Modern China*, vol. 37, no. 6, pp. 646–660.

Dagher J., Gottschalk J., and Portillo R., 2010, 'Oil windfalls in Ghana: A DSGE approach', IMF Working Paper, WP/10/116, pp. 1–36.

Daily Graphic, 2010a, 'Ghana to avoid "oil curse": Says World Bank', *Daily Graphic*, 21 December.

Daily Graphic, 2010b, 'Don't sign petroleum bill: Akufo-Addo', *Daily Graphic*, 18 December.

Daily Graphic, 2012, 'Oil production now over 105,000 bpd', *Daily Graphic*, 21 December.

Daily Guide, 2009, 'Ghana's oil and gas hijacked'. Available at: http://news.myjoyonline.com/features/200912/39125.asp (accessed 21 August 2010).

Daily Guide, 2010, '$38m for oil and gas sector', *Daily Guide*, 22 December.

Damluji M., 2013, 'The oil city in focus: The cinematic spaces of Abadan in the Anglo-Iranian Oil Company's Persian story', *Comparative Studies of South Asia, Africa and the Middle East*, vol. 33, no. 1, 75–88.

Daniel P., Keen M., and McPherson C., eds, 2010, *The Taxation of Petroleum and Minerals: Principles, Problems and Practice*, Routledge, New York.

Dantzig A.V., 1980, *Forts and Castles of Ghana*, Sedco Enterprise, Accra.

Darkwah K.A., 2013, 'Keeping hope alive: An analysis of training opportunities for Ghanaian youth in the emerging oil and gas industry in Ghana', *International Development Planning Review*, vol. 35, no. 2, pp. 119–134.

Davis J., Ossowski R., Daniel J., and Barnett S., 2001, 'Oil funds: Problems posing as solutions?' *Finance and Development*, vol. 38, no. 4, pp. 1–7.

Davis L., 2013, 'On rent', *Work and Wealth Jobs USA*, 17 April. Available at: http://workandwealth.com/rent/?utm_source=rss&utm_medium=rss&utm_campaign=rent (accessed 10 August 2013).

De Smith A.S., 1957, 'The independence of Ghana', *Modern Law Review*, vol. 20, no. 4, pp. 347–363.

De Soto H., 2000, *The Mystery of Capital: Why Capitalism Triumphs in the West and Fails Everywhere Else*, Bantam Press, New York.

De Soto H., 2011, 'This land is your land: A conversation with Hernando de Soto', *World Policy Journal*, vol. 28, summer, pp. 35–40.

De Soysa I. and Neumayer E., 2007, 'Resource wealth and the risk of civil war onset: Results from a new dataset of natural resource rents, 1970–1999', *Conflict Management and Peace Science*, vol. 24, no. 3. pp. 201–218.

DeGrassi A., 2008, '"Neopatrimonialism" and agricultural development in Africa: Contributions and limitations of a contested concept', *African Studies Review*, vol. 51, no. 3, pp. 107–133.

Demsetz H., 1967, 'Toward a theory of property rights', *American Economic Review*, vol. 57, no. 2, pp. 347–359.

Demsetz H., 2002, 'Toward a theory of property rights II: The competition between private and collective ownership', *Journal of Legal Studies*, vol. XXXI, June, pp. s653–s672.

Dickson K.B., 1965, 'Evolution of seaports in Ghana: 1800–1928', *Annals of the Association of American Geographers*, vol. 55, no. 1, pp. 98–111.

Dumett R.E., 1983, 'African merchants of the Gold Coast, 1860–1905: Dynamics of indigenous entrepreneurship', *Comparative Studies in Society and History*, vol. 25, no. 4, pp. 661–693.

Dunn B., 2007, 'Accumulation by dispossession or accumulation of capital? The case of China', *Journal of Australian Political Economy*, vol. 60, pp. 6–27.

Dunn B., 2009, *Global Political Economy*, Pluto Press, London.

Dunn B., 2012 'Skills, credentials and their unequal reward in a heterogeneous global political economy', *Journal of Sociology*, doi:10.1177/1440783312459103.

Durkheim E. (1951) *Suicide: A Study in Sociology*, Free Press, Glencoe IL.

Dye R.F. and England W.R., eds, 2009, *Land Value Taxation: Theory, Evidence, and Practice*, Lincoln Institute of Land Policy, Cambridge.

Dzokoto V.A.B., Young J., and Mensah C.E., 2010, 'A tale of two Cedis: Making sense of a new currency in Ghana', *Journal of Economic Psychology*, vol. 31, no. 4, pp. 520–526.

Economic Community of West African States, 2008, 'The ECOWAS Conflict Prevention Framework', Mediation and Security Council, ECOWAS, Ougadougou.

Edem R., 2011, 'The expected impact of the oil discovery in Takoradi on land use patterns and growth of the city', BSc. dissertation submitted to the Department of Planning, Kwame Nkrumah University of Science and Technology, Ghana.

Egiegba A., 2013, 'Have we heard the last? Oil, environmental insecurity, and the impact of the amnesty programme on the Niger Delta resistance movement', *Review of African Political Economy*, vol. 40, no. 137, pp. 447–465.

Egyir I.K., 2012, 'The impacts of oil and gas activities on fisheries in the Western Region of Ghana', Master's Degree thesis in international fisheries management submitted to the Norwegian College of Fisheries Science, University of Tromso, Norway.

Ekpenyong S., 1989, 'Housing, the state, and the poor in Port Harcourt', *Cities*, February, pp. 39–49.

Ekuful C., 2010, 'Jubilee oil gushes out today', *Ghanaian Times*, 15 December.

El-Gamal M.A. and Jaffe A.M., 2010, *Oil, Dollars, Debt, and Crises: The Global Curse of Black Gold*, Cambridge University Press, Cambridge.

Elbendak, O.E., 2008, 'Urban transformation and social change in a Libyan city: An anthropological study of Tripoli', PhD thesis submitted to the Department of Anthropology, National University of Ireland, Maynooth.

Elischer S., 2011, 'Measuring and comparing party ideology in nonindustrialized societies: Taking party manifesto research to Africa', *Democratization*, vol. 19, no. 4, pp. 642–667.

Elkind S.S., 2012, 'Oil in the city: The fall and rise of oil drilling in Los Angeles', *Journal of American History*, June, pp. 82–90.

Enin K., 2011, 'Security and crime management in the wake of the discovery and exploration of oil and gas in Ghana', Avoiding the oil curse in Ghana, International conference held between 24 and 27 November, Koforidua.

ENNIMIL, 2011, 'Takoradi market structure is fallen apart', *ENNIMIL News File*. Available at: http://ennimil.blogspot.com.au (accessed 13 July 2012).

Environmental Protection Agency (EPA), 2011a, *National Greenhouse Gas Inventory Report for 1990–2006*, EPA, Accra.

Environmental Protection Agency (EPA), 2011b, *Ghana's Second National Communication to the UNFCC*, EPA, Accra.

Environmental Protection Agency (EPA), 2011c, 'Offshore oil and gas development in Ghana', Environmental Information and Data Management Department, EPA, Accra.

Escobar A., 1998, 'Whose knowledge, whose nature? Biodiversity, conservation, and the political ecology of social movements', *Journal of Political Ecology*, vol. 5, pp. 53–82.

Essah D.S., 2008, 'Fashioning the nation: Hairdressing, professionalism and the performance of gender in Ghana, 1900–2006', PhD dissertation submitted to the History Department, University of Michigan, Michigan MI.

Essien S.M., 2011, 'Sex for sale: The rise of child prostitution in Takoradi', *Daily Guide*, 1 October. Available at: www.dailyguideghana.com/?p=28530 (accessed 17 March 2012).

Faah G., 2008, 'Road transportation impact on Ghana's future energy and environment', PhD thesis submitted to the Faculty of Economics and Business Administration, Technische Universität Bergakademie, Freiberg.

Fainstein S., 2012, 'Land value capture and justice', in Ingram G.K. and Hong Y.-H., eds, *Value Capture and Land Policies*, Lincoln Institute of Land Policy, Cambridge, pp. 21–40.

Falola T. and Genova A., 2005, *The Politics of the Global Oil Industry*, Praeger Publishers, Westport CT.

Farouk B. and Mensah O., 2012, ' "If in doubt, count": The role of community-driven enumerations in blocking eviction in Old Fadama, Accra', *Environment and Urbanization*, vol. 24, no. 1, pp. 47–57.

Farvacque-Vitkovic C., Raghunath M., Eghoff C., and Boakye C., 2008, 'Development of the cities of Ghana: Challenges, priorities and tools', Africa Region Working Paper Series, no. 110.

Feder K., 2001, 'Henry George on property rights: Reply to John Pullen', *American Journal of Economics and Sociology*, vol. 60, no. 2, pp. 565–579.

Fentiman A., 1996, 'The anthropology of oil: The impact of the oil industry on a fishing community in the Niger Delta', *Social Justice*, vol. 23, no. 4, pp. 87–99.

Ferguson J., 1990a, 'Mobile workers, modernist narratives: A critique of the historiography of transition on the Zambian Copperbelt, Part One'. *Journal of Southern African Studies*, vol. 16, no. 3, pp. 385–412.

Ferguson J., 1990b, 'Mobile workers, modernist narratives: A critique of the historiography of transition on the Zambian Copperbelt, Part Two', *Journal of Southern African Studies*, vol. 16, no. 4, pp. 603–621.

Ferguson J., 2005, 'Seeing like an oil company: Space, security, and global capital in neo-liberal Africa', *American Anthropologist*, vol. 107, no. 3, pp. 377–382.

Fortescue D., 1990, 'The Accra crowd, the asafo, and the opposition to the Municipal Corporations Ordinance, 1924–25', *Canadian Journal of African Studies/Revue Canadienne des Études Africaines*, vol. 24, no. 3, pp. 348–375.

Foster V. and Pushak N., 2011, 'Ghana's infrastructure: A continental perspective', World Bank Policy Research Working Papers, 5600, World Bank, Washington DC.

Frankel B., 2004, 'Oils ain't oils and super ain't super', *Journal of Australian Political Economy*, vol. 53, June, pp. 67–80.

Franks S. and Nunnally S., 2011, *Barbarians of Oil: How the World's Oil Addiction Threathens Global Prosperity and Four Investment to Protect Your Wealth*, John Wiley and Sons Inc., Hobokon NJ.

Frederisken E., 2007, 'Labor mobility, household production and the Dutch disease', Working Paper, University of Copenhagen, Denmark.

Freund B., 2001, 'Contrasts in urban segregation: A tale of two African cities, Durban (South Africa) and Abidjan (Cote d'Ivoire)', *Journal of Southern African Studies*, vol. 27, no. 3, pp. 527–546.

Frieden J., 2000, 'Towards a political economy of takings', *Festschrift*, vol. 3, pp. 137–147.

Friedrich-Ebert-Stiftung, 2011, *Youth and Oil and Gas: Governance in Ghana Nationwide Survey*, Friedrich-Ebert-Stiftung, Accra.

Fuccaro N., 2013, 'Introduction: Histories of oil and urban modernity in the Middle East', *Comparative Studies of South Asia, Africa and the Middle East*, vol. 33, no. 1, pp. 1–6.

Gadugah N., 2009, 'Takoradi: Tenants in danger as house owners cash in on oil money'. Available at: http://news.myjoyonline.com/business/200912/39044.asp (accessed 28 December 2009).

Gadugah N., 2010, 'Parliamentary committee rejects 10% of oil revenue demands by Western chiefs'. Available at: http://news.myjoyonline.com/news/201011/56186.asp (accessed 27 January 2011).

Gaffney M., 1994, 'Neo-classical economics as a stratagem against Henry George', in Harrison F., ed., *The Corruption of Economics*, Shepheard-Walwyn Publishing Co., London, pp. 29–164.

Gaffney M., 2008, 'Keeping land in capital theory: Ricardo, Faustmann, and George', *American Journal of Economics and Sociology*, vol. 67, no. 1, pp. 119–142.

Gaffney M., 2009, *After the Crash: Designing a Depression-Free Economy*, Wiley-Blackwell, Chichester.

García-Rodríguez F.J., García-Rodríguez J.L., Castilla-Gutiérrez C., and Major S.A., 2013, 'Corporate social responsibility of oil companies in developing countries: From altruism to business strategy', *Corporate Social Responsibility and Environmental Management*, vol. 20, no. 6, pp. 371–384.

Gary I., 2009, *Ghana's Big Test: Oil's Challenge to Democratic Development*, Isodec, Ghana and OXFAM, America.

Gatsi J., 2010, 'Oil revenue collateralisation in Ghana'. Available at: http://ssrn.com/abstract=1718551 (accessed 17 January 2014).

Gay E.F., 1930, 'Historical records', *American Economic Review*, vol. 20, no. 1, pp. 1–8.

Genugten S.V., 2011, 'Libya after Gadhafi', *Survival: Global Politics and Strategy*, vol. 53, no. 3, pp. 61–74.

George H., 1881, 'The land question: What it involves, and how alone it can be settled' [first published in 1881 as 'The Irish land question'], Schalkenbach Foundation, New York.

George H., 1898, *The Science of Political Economics*, Kegan Paul, Trench, Trubner and Co., Ltd, London.

George H., [1879] 2006, *Progress and Poverty*, Robert Schalkenbach Foundation, New York.

Ghana AIDS Commission, 2012, *Ghana Country AIDS Progress Report*, Ghana AIDS Commission, Accra.

Ghana Health Service (GHS), 2009, *2009 GHS Annual Report*, GHS, Sekondi-Takoradi.

Ghana Health Service (GHS), 2010, *Half Year Review Report of the Regional Health Directorate, Western Region*, GHS, Sekondi-Takoradi.

Ghana Maritime Authority, 2011, *Annual Report of the Ghana Maritime Authority*, Ghana Maritime Authority, Accra.

Ghana News Agency, 2008, 'Russian oil company to invest US$100m in Ghana'. Available at: www.myzongo.com/2008/06/03/russian-oil-company-to-invest-us100m-in-ghana/ (accessed 13 November 2008).

Ghana News Agency, 2008, 'Study of effects of mining on Obuasi launched'. Available at: http://news.myjoyonline.com/business/200806/16998.asp (accessed 22 June 2009).

Ghana News Agency, 2009, 'Minority cries foul over Ghana's yet-to-flow oil'. Available at: http://news.myjoyonline.com/politics/200911/37825.asp (accessed 23 August 2010).

Ghana News Agency, 2011, 'EPA develops guidelines to regulate oil and gas activities', *Ghana News Agency*, 1 February. Available at: www.ghananewsagency.org/s_science/r_25008/ (accessed 3 February 2011).

Ghana News Agency, 2012, 'IEA commends minister for finance and economic planning', *Ghana News Agency*, 17 May.

Ghana News Agency, 2013, 'Child prostitution booming at Sekondi: Pimps pay Ghc 2.00 for sex'. Available at: www.myradiogoldlive.com/index.php/general-news/1845-child-prostitution-booming-at-sekondi-pimps-pay-ghc-200-for-sex (accessed 17 March 2012).

Ghana Oil and Gas Service Providers Association (GOGSPA), 2010, 'A report on the meeting held at GOGSPA on concerns raised on medical certificates', GOGSPA, Accra, 13 September.

Ghana Railway Company Ltd, 2008, *Ghana Railway Company Limited: Profile of Railway* (in the 2008 company diary), Ghana Railway Company Ltd, Takoradi.

Ghana Real Estate Developers Association (GREDA), 2010, *Presentation by GREDA to Parliament on STX Housing*, GREDA, Accra.

Ghana Statistical Service (GSS), 1989, *1984 Population and Housing Census*, GSS, Accra.

Ghana Statistical Service (GSS), 2000, *Ghana Living Standards Survey: Report of the Fourth Round*, GSS, Accra.

Ghana Statistical Service (GSS), 2005, *2000 Population and Housing Census*, GSS, Accra.

Ghana Statistical Service (GSS), 2007, *Pattern and Trends of Poverty in Ghana, 1991–2006*, GSS, Accra.

Ghana Statistical Service (GSS), 2008, *Ghana Living Standards Survey: Report of the Fifth round*, GSS, Accra.

Ghana Statistical Service (GSS), 2012, *2010 Population and Housing Census*, GSS, Accra.

Ghana Tourism Authority, 2012, 'Western Region project registration data', Ghana Tourism Authority, Takoradi.

Ghanaian Times, 2010, 'Jubilee oil partnership commences community health outreach project', *Ghanaian Times*, 30 November.

Gilbert A. and Pasty H., 1985, *The Political Economy of Land: Urban Development in an Oil Economy*, Gower, Aldershot, UK, Brookfield VT.

Glassman J., 2006, 'Primitive accumulation, accumulation by dispossession, accumulation by "extra-economic" means', *Progress in Human Geography*, vol. 30, no. 5, pp. 608–625.

Glynn S., 2009, 'Fighting back: Lessons from 100 years of housing campaigns', in Glynn, S., ed., *Where the Other Half Lives: Lower Income Housing in a Neoliberal World*, Pluto Press, New York, pp. 281–319.

Goldsworthy D., 1973, 'Ghana's second republic: A post mortem', *African Affairs*, vol. 72, no. 286, pp. 8–25.

Goodman J., 2011, 'Responding to climate crises: "Modernisation, limits, socialism"', *Journal of Australian Political Economy*, no. 66, pp. 144–165.

Goodman J. and Worth D., 2008, 'The mineral's boom and Australia's "resource curse"', *Journal of Australian Political Economy*, no. 61, pp. 201–217

Goody J., 1968, 'The myth of a state', *Journal of Modern African Studies*, vol. 6, no. 4, pp. 461–473.

Gordon A., Pulis A., and Owusu-Adjei E., 2011, 'Smoked marine fish from Western Region, Ghana: A value chain assessment', World Fish Center, USAID Integrated Coastal and Fisheries Governance Initiative for the Western Region, Ghana.

Gough K.V. and Yankson P.W.K., 2012, 'Exploring the connections: Mining and urbanisation in Ghana', *Journal of Contemporary African Studies*, vol. 30, no. 4, pp. 651–668.

Government of Ghana, 2009, *The National Oil Spill Response Dispersant Use Policy*, Parliament of Ghana, Accra.

Government of Ghana, 2010a, *Ghana Shared Growth and Development Agenda, 2010–2013*, Government of Ghana, Accra.

Government of Ghana, 2010b, *Coordinated Programme for Economic and Social Development Policies, 2010–2016*, Government of Ghana, Accra.

Government of Ghana, 2010c, *Spillage Contingency Plan*, Parliament of Ghana, Accra.

Government of Ghana, 2012, *Overview of the Better Ghana Agenda*, Ministry of Information, Accra.

Gowdy J. and McDaniel C., 1999, 'The physical destruction of Nauru: An example of weak sustainability', *Land Economics*, vol. 75, no. 2, pp. 333–338.

Grant R., 2009, *Globalizing City: The Urban and Economic Transformation of Accra, Ghana*, Syracuse University Press, New York.

Grant R., 2012, 'Accra: A globalizing city', in Derudder B., Hoyler M., Taylor J.P., and Witlox F., eds, *International Handbook of Globalization and World Cities*, Edward Elgar, Cheltenham, pp. 455–465.

Grant R. and Thompson D., 2013, 'The development complex, rural economy and urban-spatial and economic development in Juba, South Sudan', *Local Economy*, vol. 28, March, pp. 218–230.

Greif A., 1998, 'Historical and comparative institutional analysis', *American Economic Review*, vol. 88, no. 2, pp. 80–84.

Grill C., 1984, 'Urbanisation in the Arabian Peninsula' *Occasional Papers*, ISSN 03070654.

Gyampo R.E.V., 2011, 'Saving Ghana from its oil: A critical assessment of preparations so far made', *Africa Today*, vol. 57, no. 4, pp. 49–69.

Gyampo R.E.V., 2012, 'The youth and political ideology in Ghanaian politics: The case of the fourth republic', *Africa Development*, Vol. XXXVII, no. 2, pp. 137–165.

Gyampo R.E.V., 2013, 'Youth in parliament and youth representation in Ghana', *Journal of Asian and African Studies*, doi:10.1177/0021909613511941.

Gyimah-Boadi E., 2009, 'Another step forward for Ghana', *Journal of Democracy*, vol. 20, no. 2, pp. 138–152.

Gyimah-Boadi, E. and Prempeh, H.K., 2012, 'Oil, politics, and Ghana's democracy', *Journal of Democracy*, vol. 23, no. 3, pp. 94–108.

Hackman N.A., 2009, 'Was Ghana right in choosing royalty tax system for the oil sector?' *Danquah Institute Quarterly*, October, pp. 1–24.

Hamilton D.I., 2011, 'Oil and gas companies and community crises in the Niger Delta', *American Review of Political Economy*, vol. 9, no. 1, pp. 3–17.

Hammond J.L., 2011, 'The resource curse and oil revenues in Angola and Venezuela', *Science and Society*, vol. 75, no. 3, pp. 348–378.

Harris J. and Todaro M., 1970, 'Migration, unemployment and development: A two-sector analysis', *American Economic Review*, vol. 60, no. 1, pp. 126–142.

Hart K., 1973, 'Informal income opportunities and urban employment in Ghana', *Journal of Modern African Studies*, vol. 11, no. 1, pp. 61–89.

Hart M., 2009, 'Oil "hot spot": Ghana must proceed with caution'. Available at: www.oxfamamerica.org/newsandpublications/press_releases/oil-hot-spot-ghana-must-proceed-with-caution (accessed 9 June 2009).

Harvey D., 2003, *The New Imperialism*, Oxford University Press, Oxford.

Harvey D., 2006a, *Limits to Capital*, Verso, London and New York.

Harvey D., 2006b, *Spaces of Global Capitalism: Towards a Theory of Uneven Geographical Development*, Verso, New York.

Harvey D., 2008, 'The right to the city', *New Left Review*, vol. 53, September, pp. 23–40.

Harvey D., 2011, 'The future of the commons', *Radical History Review*, no. 109, winter, pp. 101–107.

Haynes J., 1991, 'Railway workers and the P.N.D.C. Government in Ghana, 1982–90', *Journal of Modern African Studies*, vol. 29, no. 1, pp. 137–154.

Heller P.R.P., 2013, 'Civil society and the evolution of accountability in the petroleum sector', in Appiah-Adu K., ed., *Governance of the Petroleum Sector in an Emerging Developing Economy*, Gower Applied Research, Ashgate, Surrey and Burlington VT, pp. 89–107.

Herbert-Cheshire L. and Lawrence G., 2002, 'Political economy and the challenge of governance', *Journal of Australian Political Economy*, no. 50, December, pp. 137–145.

Hill P., 1966, 'A plea for indigenous economics: The West African example', *Economic Development and Cultural Change*, vol. 15, no. 1, pp. 10–20.

Hilling D., 1969, 'The evolution of the new ports of West Africa', *Geographical Journal*, vol. 135, no. 3, pp. 365–378.

Hilling D., 1975, 'Port specialization and efficiency: The case of Ghana', *Maritime Studies and Management*, vol. 3, no. 1, pp. 13–20.

Hilson G., 2002, 'Harvesting mineral riches: 1000 years of gold mining in Ghana', *Resources Policy*, vol. 28, pp. 13–26.

Hilson G. and Banchirigah M., 2009, 'Are alternative livelihood projects alleviating poverty in mining communities? Experiences from Ghana', *Journal of Development Studies*, vol. 45, no. 2, pp. 172–196.

Howard R., 1978, *Colonialism and Underdevelopment in Ghana*, Croom Helm, London.

Hubacek K. and Bergh C., 2006, 'Changing concepts of "land" in economic theory: From single to multidisciplinary approaches', *Ecological Economics*, vol. 56, no. 1, pp. 5–27.

Hubbard P., 2012, 'Afterword: Exiting Amsterdam's red light district', *City*, vol. 16, no. 1–2, pp. 195–201.

Hubbard P., Boydell S., Crofts P., Prior J.H., and Searle G.H., 2013, 'Noxious neighbours? Interrogating the impacts of sex premises in residential areas', *Environment and Planning A*, vol. 45, pp. 126–141.

Huber M., 2013, 'Fueling capitalism: Oil, the regulation approach, and the ecology of capital', *Economic Geography*, vol. 89, no. 2, pp. 171–194.

Hufstader C., 2008, 'The coming oil boom in Ghana'. Available at: www.publishwhat youpay.org/en/resources/coming-oil-boom-ghana (accessed 29 October 2008).

Hutchful E., 1979, 'A tale of two regimes: Imperialism, the military and class in Ghana', *Review of African Political Economy*, no. 14, January to April, pp. 36–55.

Idemudia U., 2007, *Corporate Partnerships and Community Development in the Nigerian Oil Industry*, UNRISD, Geneva.

Ifeka C., 2004, 'Violence, market forces and the militarisation of the Niger Delta', *Review of African Political Economy*, vol. 39, no. 99, pp. 144–150.

IMANI, 2012, 'IMANI alert: Sekondi industrial estate may endanger part of CDB loan', Press release, 8 January.

Imobighe M.D., 2011, 'Paradox of oil wealth in the Niger-Delta region of Nigeria: How sustainable is it for national development?' *Journal of Sustainable Development*, vol. 4, no. 6, pp. 160–168.

Ingram G.K. and Hong Y.-H., 2012, 'Land value capture: Types and outcomes', in Ingram G.K. and Hong Y.-H., ed., *Value Capture and Land Policies*, Lincoln Institute, Cambridge, pp. 5–18.

Israel A.M., 1987, 'Measuring the war experience: Ghanaian soldiers in World War II', *Journal of Modern African Studies*, vol. 25, no. 1, pp. 159–168.

Jackson A., 2006, *The British Empire and the Second World War*, Hambledon Continuum, New York.

James A. and Aadland D., 2011, 'The curse of natural resources: An empirical investigation of U.S. counties', *Resource and Energy Economics*, vol. 33, no. 2, pp. 440–453.

Jang H.S., 2010, 'Social identities of Indigenous people in their twenties in contemporary Australian society based on the rural Indigenous community at Yarrabah in Queensland', Doctoral thesis, University of Sydney, Sydney.

Jedwab R. and Moradi A., 2011, 'Transportation infrastructure and development in Ghana', Working Paper No. 2011–24, Paris School of Economics.

Jeffries R., 1975, 'The labour aristocracy? Ghana case study', *Review of African Political Economy*, vol. 2, no. 3, pp. 59–70.

Jeffries R., 1978, *Class, Power and Ideology in Ghana: The Railwaymen of Sekondi*, Cambridge University Press, Cambridge.

Jeffreys E., 2003, 'Transnational prostitution: A global feminist response?' *Australian Feminist Studies*, vol. 18, no. 41, pp. 211–216.

Jenkins P., Robson P., and Cain A., 2002, 'Local responses to globalization and peripheralization in Luanda, Angola', *Environment and Urbanization*, vol. 14, no. 1, pp. 115–127.

Jessop B., 1990, *State Theory: Putting the Capitalist State in Its Place*, Cambridge University Press, Cambridge.

Jike T., 2004, 'Environmental degradation, social disequilibrium, and the dilemma of sustainable development in the Niger Delta of Nigeria', *Journal of Black Studies*, vol. 34, pp. 686–701.

Jønsson J.B. and Fold N., 2011, 'Mining from below: Taking Africa's artisanal miners seriously', *Geography Compass*, vol. 5, no. 7, pp. 479–493.

Joy Business, 2009, 'Jubilee fields caught up in turf war as GNPC, others firm interest'. Available at: http://news.myjoyonline.com/business/200910/36401.asp (accessed 23 August 2010).

Judge D., Stoker G., and Wolman H., 1995, *Theories of Urban Politics*, Sage Publications Ltd, London.

Kapela J.M., 2009, 'Ghana's new oil: Cause for jubilation or prelude to the resource curse', Masters project submitted in partial fulfilment of the requirements for the Master of Environmental Management degree in the Nicholas School of the Environment of Duke University, Durham NC.

Kapoor I., 2004, 'Hyper-self-reflexive development? Spivak on representing the Third World "Other" ', *Third World Quarterly*, vol. 25, no. 4, pp. 627–647.

Kar K., 2005, *Practical Guide to Triggering Community-led Sanitation*, Institute of Development Studies, Brighton.

Karl L., 1997, *The Paradox of Plenty*, University of California Press, Berkeley CA.

Karl T.L., 2004, 'Oil-led development: Social, political, and economic consequences', *Encyclopedia of Energy*, vol. 4, pp. 661–672.

Kasanga K., 2001, 'Land resource management for agricultural development in Ghana', *Our Common Estate*, pp. 5–30.

Kayoke E., 2004, 'Pollutant emissions measured: Rising transport pollution in the Accra–Tema metropolitan area, Ghana', MSc thesis, Lund University, Sweden.

Keen S., 2003, 'Madness in their method', in Stilwell F. and Argyrous G., eds, *Economics as a Social Science*, Pluto Press, Melbourne, pp. 140–145.

Keizeiri K.S., 1983, 'Urbanization trends and state intervention in Libya', *Journal of Environmental Planning and Management*, vol. 27, no. 1, pp. 17–21.

Kenamore J., 1983, 'Review of oil booms: Social change in five Texas towns', *Public Historian*, vol. 5, no. 4, pp. 117–119.

Kent A. and Ikgopoleng H., 2011, 'City profile: Gaborone', *Cities*, vol. 28, no. 5, pp. 478 494.

Keynes J.M., 1936, *The General Theory of Employment, Interest and Money*, Macmillan, London.

Keynes J.M., 1973, *The Collected Writings of John Maynard Keynes: The General Theory of Employment, Interest and Money*, vol. II, Macmillan, Cambridge University Press, London.

Kimble D., 1963, *A Political History of Ghana: The Rise of Gold Coast Nationalism 1850–1928*, Clarendon Press, Oxford.

King R., 2009, 'An institutional analysis of the resource curse in Africa: Lessons for Ghana', *Consilience – The Journal of Sustainable Development*, no. 2, pp. 1–22.

Kirchherr E.C., 1968, 'Tema 1951–1962: The evolution of a planned city in West Africa', *Urban Studies*, vol. 5, no. 2, pp. 207–217.

Klaeger G., 2013, 'Dwelling on the road: Routines, rituals and roadblocks in southern Ghana', *Africa: The Journal of the International African Institute*, vol. 83, no. 3, pp. 446–469.

Klein N., 2007, *The Shock Doctrine: The Rise of Disaster Capitalism*, Penguin Books Ltd, London.

Kleist N., 2011, 'Modern chiefs: Tradition, development and return among traditional authorities in Ghana', *African Affairs*, vol. 110, no. 441, pp. 629–647.

Kokutse F., 2012, 'Ghana: Oil revives prostitution'. Available at: http://ghanaoilonline. org/2012/02/ghana-oil-revives-prostitution/ (accessed 17 March 2012).

Kolstad I. and Wiig A., 2012, 'Testing the pearl hypothesis: Natural resources and trust', *Resources Policy*, May.

Kombat A., 2013, 'Economic assessment of environmental taxes and standards in managing the environmental problems that emanate from oil and gas production on Ghana's Jubilee field', *Journal of International Real Estate and Construction Studies*, vol. 3, no. 1.

Koomson F., 2010, 'Ivory Coast oil find claim has serious implications'. Available at: http://news.myjoyonline.com/business/201003/43007 (accessed 24 August 2010).

Kopiński D., Polus A., and Tycholiz W., 2013, 'Resource curse or resource disease? Oil in Ghana', *African Affairs*, vol. 112, no. 449, pp. 583–601.

Korboe D. and Tipple G., 1995, 'City profile, Kumasi', *Cities*, vol. 12, no. 4, pp. 267–274.

KOSMOS Energy, 2010, 'KOSMOS Energy reaches settlement agreement with Ghanaian government and Ghana National Petroleum Corporation', News Release, Accra, 20 December.

Krugman P., 2011, 'The new economic geography, now middle-aged', *Regional Studies*, vol. 45, no. 1, pp. 1–7.

Kwettey F., 2010, 'Government will not collateralise 100% of the oil revenue'. Available at: http://news.myjoyonline.com/news/201012/57040.asp (accessed 24 January 2011) (contribution to Joy FM Newsfile programme on 4 December; also reported by D. Adogla).

Lacher W., 2011, 'Families, tribes, and cities in the Libyan Revolution', *Middle Eastern Policy*, vol. 18, no. 4, pp. 140–154.

Larbi W., Antwi A., and Olomolaiye P., 2004, 'Compulsory land acquisition in Ghana: Policy and praxis', *Land Use Policy*, vol. 21, pp. 115–127.

Larsen E., 2005, 'Are rich countries immune to the resource curse? Evidence from Norway's management of its oil riches', *Resource Policy*, vol. 30, pp. 75–86.

Larsen E., 2006, 'Escaping the resource curse and the Dutch disease? When and why Norway caught up with and forged ahead of its neighbours', *American Journal of Economics and Sociology*, vol. 65, no. 3, pp. 605–640.

Laumann D.H., 1993, '"Compradore-in-arms": The Fante confederation project (1868–1872)', *Ufahamu: A Journal of African Studies*, vol. 21, no. 1–2, pp. 120–136.

Lawrie M., Tonts M., and Plummer P., 2011, 'Boomtowns, resource dependence and socio-economic well-being', *Australian Geographer*, vol. 42, no. 2, pp. 139–164.

Le T.-H. and Chang Y., 2012, 'Oil price shocks and gold returns', *International Economics*, vol. 131, March, pp. 71–103.

Lesourne J. and Ramsay W., 2009, *Governance of Oil in Africa: Unfinished Business*, Ifri, Paris.

Linder P., 2006, 'Speech by H.E. Peter Linder, German Ambassador to Ghana', delivered on the occasion of the official launch of the Mercedes S-Class in Accra, Available at: www.accra.diplo.de/Vertretung/accra/de/01/2006_Reden_PressReleases/mercedes_S_ Klasse.html (accessed 26 September 2009).

Lippit V., 2011, 'Introduction: China's rise in the global economy', *Review of Radical Political Economics*, vol. 43, no. 1, pp. 5–8.

Loomer C.W., 1951, 'A comment on professor Schultz' framework for land economics', *Journal of Farm Economics*, vol. 33, no. 3, pp. 389–396.

Luong P.J. and Weinthal E., 2010, *Oil is not a Curse: Ownership Structure and Institutions in Soviet Successor States*, Cambridge University Press, Cambridge.

McCaskie T, 2008, 'The United States, Ghana and oil: Global and local perspectives', *African Affairs*, vol. 107, no. 428, pp. 313–332.

Maconachie R., Tanko A., and Zakariya M., 2009, 'Descending the energy ladder? Oil price shocks and domestic fuel choices in Kano, Nigeria', *Land Use Policy*, vol. 26, no. 4, pp. 1090–1099.

Mahama C, 2006, 'Land and property markets in Ghana', Discussion paper prepared by the Royal Institution of Chartered Surveyors for presentation at the 2006 World Urban Forum by Callistus Mahama (Kwame Nkrumah University of Science and Technology, Ghana) and Adarkwah Antwi (University of Wolverhampton, England).

Mahama J., 2013, 'State of the nation address', National address given by the president at the Ghana Parliament House, Accra, 21 February.

Mahama F., 2012, 'Study of vehicular traffic congestion in the Sekondi-Takoradi metropolis', Master's dissertation, Department of Mathematics, Kumasi, KNUST.

Mail, 2011, 'Tullow Ghana presents textbooks to schools', *The Mail*, 19 January.

Malecki E.J. and Ewers M.C., 2007, 'Labor migration to world cities: With a research agenda for the Arab Gulf', *Progress in Development Studies*, vol. 31, no. 4, pp. 467–488.

Mahdavy H., 1970, 'The patterns and problems of economic development in rentier states: The case of Iran', in Cook M.A., ed., *Studies in the Economic History of the Middle East*, Oxford University Press, Oxford, pp. 428–467.

Manu D.A.K., 2011, 'The emerging oil industry in Ghana: Socio-economic and environmental impact on the people of Cape Three Points', Master's thesis in International Environmental and Development Studies, Norwegian University of Life Sciences, Norway.

Marcuzzo M.C., 2008, 'Is history of economic thought a "serious" subject?' *Erasmus Journal of Philosophy and Economics*, vol. 1, no. 1, pp. 107–123.

Marfo A.K., 2008, 'SAEMA is now STMA' Available at: http://kwameasiedumarfosstories.blogspot.com.au/2008/04/saema-is-now-stma.html on March 25 (accessed 9 June 2012).

Marful-Sau S., 2010, 'Is Ghana prepared to manage the potential environmental challenges of an oil and gas industry?' Paper presented at the Centre for Energy, Petroleum and Mineral Law and Policy, University of Dundee, Dundee.

Markusen A., 1978, 'Class, rent, and sectorial conflict: Uneven development in Western U.S. boomtowns', *Review of Radical Political Economics*, vol. 10, no. 3, pp. 117–129.

Marx K., 1956, *Capital*, vol. 3. Available at: Marx.org 1996, Marxists.org 1999. Transcribed for the internet in 1996 by Hinrich Kuhls and Zodiac, and by Tim Delaney and M. Griffin in 1999.

Marx K., [1976] 1990, *Capital, Volume 1*, Penguin Books Ltd, London.

Matter S., 2010, 'Clashing claims: Neopatrimonial governance, land tenure transformation, and violence at Enoosupukia, Kenya', *PoLAR*, vol. 33, no. 1, pp. 67–88.

Mendelson S., Cowlishaw G., and Rowcliffe J.M., 2003, 'Anatomy of a bush meat commodity chain in Takoradi, Ghana', *Journal of Peasant Studies*, vol. 31, no. 1, pp. 73–100.

Mikkelsen A., Engen O., Steineke J., Jøsendal K., and Grønhaug K., 2005, 'Consequences of critical events for the social construction of corporate social responsibility: The case of oil and gas companies in Norway', Rapport RF – 2005/193.

Ministry of Education, 1991, *History for Senior Secondary School*, Ministry of Education, Accra.

Ministry of Energy, 2010, *Local Content and Local Participation in Petroleum Activities Policy Framework*, Ministry of Energy, Accra.

Ministry of Finance and Economic Planning (MOFEP), 2010, *2011 Financial Year Budget of Ghana*, MOFEP, Accra.

Ministry of Finance and Economic Planning (MOFEP), 2011, *2012 Budget Statement: Infrastructural Development for Accelerated Growth and Job Creation*, MOFEP, Accra.

Ministry of Finance and Economic Planning (MOFEP), 2012, *2013 Budget Statement: Infrastructural Development for Accelerated Growth and Job Creation*, MOFEP, Accra.

Ministry of Finance and Economic Planning (MOFEP), 2013, *2014 Budget Statement: Sustaining Confidence in the Future of the Ghanaian Economy*, MOFEP, Accra.

Ministry of Food and Agriculture, 2011, 'Revised premix fuel guidelines: Guidelines for the re-organisation of premix fuel allocation, distribution and sale', Fisheries Commission, Accra.

Ministry of Information, 2012, *Overview of the Better Ghana Agenda*, Ministry of Information, Accra.

Ministry of Local Government and Rural Development, 2010, *National Urban Policy*, Ministry of Local Government and Rural Development, Accra.

Mo Ibrahim Foundation, 2012a, *Ibrahim Index of African Governance Data Report*, Mo Ibrahim Foundation, London.

Mo Ibrahim Foundation, 2012b, *Ibrahim Index of African Governance: Summary*, Mo Ibrahim Foundation, London.

Mohammed A.-N., 2013, 'Fiscal regime analysis: What lessons will Ghana learn from Norway', *CEPMLP Annual Review – CAR*, vol. 16, pp. I–III and 1–25.

Mohan G., 1996, 'Adjustment and decentralization in Ghana: A case of diminished sovereignty', *Political Geography*, vol. 15, no. 1, pp. 75–94.

Mohan G., 2013, ' "The Chinese just come and do it": China in Africa and the prospects for development planning', *International Development Planning Review*, vol. 35, no. 3, pp. v–xxi.

Mohan G. and Power M., 2009, 'Africa, China and the "new" economic geography of development', *Singapore Journal of Tropical Geography*, vol. 30, pp. 24–28.

Mohan J., 1966, 'Varieties of African socialism', *Socialist Register*, pp. 220–266.

Morris A., 2009, 'Living on the margins: Comparing older private renters and older public housing tenants in Sydney, Australia', *Housing Studies*, vol. 24, no. 5, pp. 693–707.

Moss T. and Young L., 2009, 'Saving Ghana from its oil: The case for direct cash distribution', Centre for Global Development Working Paper no. 186, Washington DC.

Myers K., 2010, 'Selling oil assets in Uganda and Ghana: A taxing problem', Revenue Watch Institute Release, New York, 16 August.

National Development Planning Commission (NDPC), 2011, 'Avoiding the oil curse in Ghana', International conference held between 24 and 27 November, Koforidua.

Nations Online Project, 2013, 'Ghana map'. Available at: www.nationsonline.org/maps/ghana_map.jpg (accessed 4 November 2013).

Naudé W., 2011, 'Economic development in Sub-Saharan Africa: The case of the big four', Maastricht School of Management Working Paper No. 2011/34, Maastricht.

Ndjio B., 2009, 'Shanghai beauties and African desires: Migration, trade and Chinese prostitution in Cameroon', *European Journal of Development Research*, vol. 21, no. 4, pp. 606–621.

New Crusading Guide, 2010a, 'How Mills is playing the oil game'. Available at: http://news.myjoyonline.com/business/201001/41092.asp (accessed 24 August 2010).

New Crusading Guide, 2010b, 'Minister "kills" oil pact'. Available at: http://news.myjoy online.com/news/201001/40913.asp (accessed 24 August 2010).

New Crusading Guide, 2013, 'A trip down memory lane … Kan Dapaah VRS Tsatsu Tsikata in a battle over drill ships, oil rigs & "hedging dealings"!' *New Crusading Guide*, 16 September.

New Patriotic Party (NPP), 2012, *Manifesto*, NPP, Accra.

Newman P., 1991, 'Cities and oil dependence', *Cities*, August, pp. 170–173.

Nichols J.C., 1914, 'Housing and the real estate problem', *Annals of the American Academy of Political and Social Science*, vol. 51, pp. 132–139.

Njoh A.J., 2012, *Urban Planning and Public Health in Africa: Historical, Theoretical and Practical Dimensions of a Continent's Water and Sanitation Problematic*, Ashgate, Aldershot and Burlington VT.

Nketsiah N.K., 2013, 'Africa culture in governance and development: The Ghana paradigm', Unpublished manuscript.

Nkrumah K., 1962, 'Ghana: Law in Africa', Speech made by President Nkrumah at the formal opening of the Accra Conference on Legal Education and of the Ghana Law School, 4 January 1962, *Journal of African Law*, vol. 6, no. 2, pp. 103–112.

Nuamah S., 2013, 'Ghanaians abroad urged to support nation's dev.', *Ghanaian Times*, 26 January.

Nukunya G.K., 2011, *Tradition and Change in Ghana: An Introduction to Sociology*, 2nd edn, Ghana Universities Press, Accra.

Nwoke C.N., 1984, 'World mining rent: An extension of Marx's theories', *Review*, vol. 8, no. 1, pp. 29–89.

Nwoke C.N., 1986, 'Towards authentic economic nationalism in Nigeria', *Africa Today*, vol. 33, no. 4, pp. 51–69.

Nyamnjoh F., 2012, '"Potted plants in greenhouses": A critical reflection on the resilience of colonial education in Africa', *Journal of Asian and African Affairs*, vol. 47, no. 2, pp. 129–154.

Nyanor C.O., 2000, Contribution to the debate at the consideration state of the Internal Revenue Bill, 17 October, Parliamentary Debates: Official Report, Parliament of Ghana, Accra.

Nyarko E., Botwe O.B., Lamptey E., Nuotuo O., Foli B.A., Addo M.A., 2011, 'Toxic metal concentrations in deep-sea sediments from the Jubilee oil field and surrounding areas off the western coast of Ghana', *Tropical Environmental Research*, vols 9 and 10, pp. 584–595.

O'Connor J., 1988, 'Capitalism, nature, socialism a theoretical introduction', *Capitalism Nature Socialism*, vol. 1, no. 1, pp. 11–38.

O'Connor J., 1994, 'Is sustainable capitalism possible?' in O'Connor M., ed., *Is Capitalism Sustainable? Political Economy and the Politics of Ecology*, Guilford Press, New York, pp. 152–175.

O'Connor J., 2002, *The Fiscal Crises of the State*, Transaction Publishers, New Brunswick, NJ.

O'Faircheallaigh C., 2008, 'Negotiating cultural heritage? Aboriginal–mining company agreements in Australia', *Development and Change*, vol. 39, no. 1, pp. 25–51.

O'Faircheallaigh C., 2009, 'Effectiveness in social impact assessment: Aboriginal peoples and resource development in Australia', *Impact Assessment and Project Appraisal*, vol. 27, no. 2, pp. 95–110.

O'Faircheallaigh C., 2010, 'Public participation and environmental impact assessment: Purposes, implications, and lessons for public policy making', *Environmental Impact Assessment Review*, no. 30, pp. 19–27.

O'Faircheallaigh C., 2013, 'Extractive industries and indigenous peoples: A changing dynamic?' *Journal of Rural Studies*, vol. 30, pp. 20–30.

Obeng-Odoom F., 2009, 'Has the habitat for humanity housing scheme achieved its goals? A Ghanaian case study', *Journal of Housing and the Built Environment*, vol. 24, no. 1, pp. 67–84.

Obeng-Odoom F., 2010, 'Urban real estate in Ghana: A study of housing-related remittances from Australia', *Housing Studies*, vol. 25, no. 3, pp. 357–373.

Obeng-Odoom F., 2011, 'Real estate agents in Ghana: A suitable case for regulation?' *Regional Studies*, vol. 45, no. 3, pp. 403–416.

Obeng-Odoom F., 2012a, 'On the origin, meaning, and evaluation of urban governance', *Norwegian Journal of Geography*, vol. 66, pp. 204–212.

Obeng-Odoom F., 2012b, 'Good property valuation in emerging real estate markets? Evidence from Ghana', *Surveying and Built Environment*, vol. 22, November, pp. 37–60.

Obeng-Odoom F., 2012c, 'Neoliberalism and the urban economy in Ghana: Urban employment, inequality, and poverty', *Growth and Change: A Journal of Regional and Urban Policy*, vol. 43, no. 1, pp. 85–109.

Obeng-Odoom F., 2013a, 'Ideology in the 2012 elections in Ghana', *Journal of African Elections*, vol. 12, no. 2, pp. 75–95.

Obeng-Odoom F., 2013b, *Governance for Pro-Poor Urban Development: Lessons from Ghana*, Routledge, London.

Obeng-Odoom F., 2013c, 'Managing land for the common good? Evidence from a community development project at Agona, Ghana', *Journal of Pro-Poor Growth*, vol. 1, no. 1, pp. 29–46.

Obeng-Odoom F. and Amedzro L., 2011, 'Nezadostno število in neustrezna kakovost stanovanj v Gani' [Inadequate housing in Ghana], *Urbani izziv*, vol. 22, no. 1, pp. 49–59 [127–137].

Obeng-Odoom F. and Ameyaw S., 2011, 'The state of surveying in Africa: A Ghanaian perspective', *Property Management*, vol. 29, no. 3, pp. 262–284.

Obi C., 1997, 'Globalisation and local resistance: The case of the Ogoni versus Shell', *New Political Economy*, vol. 2, no. 1, pp. 137–148.

Obi C., 2007, 'Oil and development in Africa: Some lessons from the oil factor in Nigeria for the Sudan', *DIISS Report*, vol. 8, pp. 9–34.

Obi C., 2008, 'Enter the dragon? Chinese oil companies and resistance in the Niger Delta', *Review of African Political Economy*, vol. 35, no. 117, pp. 417–434.

Obi C., 2009, 'Nigeria's Niger Delta: Understanding the complex drivers of violent oil-related conflict', *Africa Development*, vol. XXXIV, no. 2, pp. 103–128.

Obi C., 2010, 'Oil extraction, dispossession, resistance, and conflict in Nigeria's oil-rich Niger Delta', *Canadian Journal of Development Studies*, vol. 30, no. 1–2, pp. 219–236.

Obimpeh S., 2000, Contribution to the debate at the consideration state of the Internal Revenue Bill, 17 October, Parliamentary Debates: Official Report, Parliament of Ghana, Accra.

Obosu-Mensah K., 1999, *Food Production in Urban Areas: A Study of Urban Agriculture in Accra, Ghana*, Ashgate, Aldershot.

Odell P., 1997, 'The global oil industry: The location of production – Middle East domination or regionalization', *Regional Studies*, vol. 31, no. 3, pp. 311–322.

Odoi-Larbi S., 2012, '$30m 5-star hotel built in Takoradi', *Ghanaian Chronicle*, 21 March.

Odoi-Larbi S., 2013, 'Ghana: Government fails to consider interest of Ghana before signing three billion Chinese loan', *Ghanaian Chronicle*, 28 May.

Offshore, 2011, 'Ghana's number one oil sample catcher', *Offshore*, June.

Offshore Ghana, 2011, '…that 10 percent imbroglio', *Offshore Ghana*, June.

Ofosu-Mensah E., 2011, 'Gold mining and the socio-economic development of Obuasi in Adanse', *African Journal of History and Culture*, vol. 3, no. 4, pp. 54–64.

Ohemeng F., 2005, 'Getting the state right: Think tanks and the dissemination of new public management ideas in Ghana', *Journal of Modern African Studies*, vol. 43, no. 3, pp. 443–465.

Ohemeng F.L.K. and Owusu F.Y., 2013, 'Implementing a revenue authority model of tax administration in Ghana: An organizational learning perspective', *American Review of Public Administration*, vol. XX, no. X, pp. 1–22.

Oil City Magazine, 2012a, 'Ghana's gold', vol. 3, no. 1.

Oil City Magazine, 2012b, 'The chocolate ingredient', vol. 3, no. 2.

Oil City Magazine, 2012c, 'Chieftaincy', vol. 3, no. 3.

Okechukwu U., Ukiwo U.O., and Ibaba I.S., eds, 2012, *Natural Resources, Conflict, and Sustainable Development: Lessons from the Niger Delta*, Routledge, London.

Okonta I. and Douglas O., 2003, *Where Vultures Feast: Shell, Human Rights and Oil*, Verso, London.

Okpanachi E. and Andrews N., 2012, 'Preventing the oil "resource curse" in Ghana: Lessons from Nigeria', *World Futures: The Journal of Global Education*, vol. 68, no. 6, pp. 430–450.

Oluduro O. and Oluduro O.F., 2012, 'Nigeria: In search of sustainable peace in the Niger Delta through the amnesty programme', *Journal of Sustainable Development*, vol. 5, no. 7, pp. 48–61.

Operations Evaluation Department, 2003, *Project performance assessment report, Ghana: Mining Sector Rehabilitation Project (Credit 1921-GH) and Mining Sector Development and Environment Project (Credit 2743-GH)*, World Bank, Washington DC. Available at: http://lnweb90.worldbank.org/OED/oeddoclib.nsf/DocUNIDView ForJavaSearch/A89AEDB05623FD6085256E37005CD815/$file/PPAR_26197.pdf (accessed 9 June 2009).

Opoku F., 2012, 'Rent tax assessment and challenges: A case study of the Sekondi-Takoradi Metropolis', Unpublished report submitted to the Ghana Institution of Surveyors as part of the requirements for acceptance into professional membership.

Oppermann M., 2000, 'Triangulation: A methodological discussion', *International Journal of Tourism Research*, vol. 2, pp. 141–146.

Osei K., 2009, 'Ghana: Why country should not bank on oil', *Public Agenda*. Available at: http://allafrica.com/stories/200901230677.html (accessed 9 June 2009).

Osei R.D. and Quartey P., 2005, 'Tax reforms and tax administration in Ghana', WIDER Research Paper 2005/10, Helsinki, Finland.

Ostrom E., 1990, *Governing the Commons: The Evolution of Institutions for Collective Action*, Cambridge University Press, New York.

Oteng-Adjei J., 2010, 'Press statement by the Ministry of Energy on request of consent to the proposed sales and purchase agreement between KOSMOS Energy Ghana HC and ExxonMobil Exploration and Production Ghana Limited', 7 August.

Owusu G. and Afutu-Kotey R., 2010, 'Poor urban communities and municipal interface in Ghana: A case study of Accra and Sekondi-Takoradi metropolis', *African Studies Quarterly*, vol. 12, no. 1, pp. 1–16.

Owusu H.J., 1998, 'Adjustment, industrial locational incentives, and structural transformation in Ghana', *East African Geographical Review*, vol. 20, no. 2, pp. 1–24.

Owusu K., 2010, 'Design and production of suggestion box for the various departments in Takoradi Polytechnic', Report submitted to the Department of Graphics, Takoradi Polytechnic, in partial fulfilment of the requirements for the award of the HND in Commercial Art, Takoradi, Ghana.

Owusu-Agyemang H., 2000, Contribution to the debate at the consideration state of the Internal Revenue Bill, 17 October, Parliamentary Debates: Official Report, Parliament of Ghana, Accra.

Owusu-Ansah A., 2012, 'Dynamics of residential property values in developing markets: A case study of Kumasi, Ghana', *Journal of International Real Estate and Construction Studies*, vol. 2, nos 1 and 2, pp. 19–35.

Owusu-Ansah D. and McFarland D.M., 1995, *Historical Dictionary of Ghana*, 2nd edn, Scarecrow Press, Inc., Metuchen NJ and London.

Owusu-Dabo E., Lewis S., McNeill A., Anderson S., Gilmore A., and Britton J., 2009, 'Smoking in Ghana: A review of tobacco industry activity', *Tobacco Control*, vol. 18, pp. 206–211.

Owusu-Koranteng D., 2008, 'Mining investment and community struggles', *Review of African Political Economy*, no. 117, pp. 467–521.

Owusuaa D., 2012, 'Gender and informality in the construction industry in Ghana's oil city Takoradi', Master's thesis, Department of Geography, Bergen.

Oyono P.R., Ribot J.C., and Larson A.M., 2006, *Green and Black Gold in Rural Cameroon: Natural Resources for Local Governance, Justice and Sustainability*, World Resources Institute, Washington DC.

Pagis M., 2009, 'Embodied self-reflexivity', *Social Psychology*, vol. 72, no. 3, pp. 265–283.

Pamford K., 2010, 'The crucial roles of Ghana's model petroleum agreement: The public policy requirements and implications', *Ghana Policy Journal*, vol. 4, December, pp. 81–95.

Panitch L., 2002, 'The impoverishment of state theory', in Arronowitz S. and Bratsis P., eds, *State Theory Reconsidered: Paradigm Lost*, University of Minnesota Press, Minneapolis MN and London, pp. 89–104.

Park A.E.W., 1965, 'Ghana: The estate duty act', *Journal of African Law*, vol. 9, no. 3, pp. 162–165.

Parker, J., 2005, 'Next door to Vichy: Soldiers, airmen, spies, and whisperers: The Gold Coast in World War II', *Journal of African History*, vol. 46, no. 3, pp. 529–531.

Patel S., 2006, '"Islands of understanding": Environmental journalism in the South Pacific', *Pacific Journalism Review*, vol. 12, no. 2, pp. 148–153.

Patton M.Q., 2002, *Qualitative Research and Evaluation Methods*, 3rd edn, Sage, London.

Pearce A. and Stilwell F., 2008, '"Green-collar" jobs: Employment impacts of climate change policies', *Journal of Australian Political Economy*, vol. 62, December, pp. 120–138.

Pegg S., 2006, 'Can policy intervention beat the resource curse? Evidence from the Chad-Cameroon pipeline project', *African Affairs*, vol. 105, no. 418, pp. 1–25.

Pejovich S., 1972, 'Towards an economic theory of the creation and specification of property rights', *Review of Social Economy*, vol. xxx, no. 3, pp. 309–325.

Pellow D. and Chazan N., 1986, *Ghana: Coping with Uncertainty*, Westview Press, Boulder CO.

Phibbs P. and Young P., 2009, 'Going once, going twice: A short history of public housing in Australia', in Glynn S., ed., *Where the Other Half Lives: Lower Income Housing in a Neoliberal World*, Pluto Press, New York, pp. 217–231.

Pickvance C., 1995, 'Marxist theories of urban politics', in Judge D., Stoker G., and Wolman H., eds, *Theories of Urban Politics*, Sage Publications, London, pp. 253–275.

Pieterse E., 2011, 'Recasting urban sustainability in the South', *Development*, vol. 54, no. 3, pp. 309–316.

Plageman N., 2008, 'Everybody likes Saturday night: A social history of popular music and masculinities in urban Gold Coast/Ghana, c.1900–1970', PhD dissertation (in the Department of History) submitted to the Graduate School, University of Indiana, Bloomington IN.

Plageman N., 2012, *Highlife Saturday Night: Popular Music and Social Change in Urban Ghana*, Indiana University Press, Bloomington IN.

Plageman N., 2013, 'Colonial ambition, common sense: Thinking, and the making of Takoradi Harbor, Gold Coast', *History in Africa*, vol. 40, pp. 317–352.

Poku-Boansi M. and Adarkwa K.K., 2011, 'An analysis of the supply of urban public transport services in Kumasi, Ghana', *Journal of Sustainable Development in Africa*, vol. 13, no. 2, pp. 24–40.

Polanyi K., [1944] 2001, *The Great Transformation: The Political and Economic Origins of Our Time*, Beacon Press, Boston MA.

Pollard J., Mcewan C., and Hughes A., eds, 2011, *Postcolonial Economies*, Zed, London.

Poteete A., 2007, 'How national political competition affects natural resource policy: The case of community-based natural resource management in Botswana', Paper presented at the 2007 Annual Conference of the Canadian Political Science Association, Saskatoon, SK, Canada, 30 May to 1 June.

Potts D., 2012a, 'Whatever happened to Africa's rapid urbanisation?' *Africa Research Institute Counterpoints Series*, February.

Potts D., 2012b, 'What do we know about urbanisation in sub-Saharan Africa and does it matter?' *International Development Review*, vol. 34, no. 1, pp. v–xxi.

Pressman S., 2001, 'State and government', in O'Hara P., ed., *Encyclopedia of Political Economy*, vol. 2, Routledge, London and New York, pp. 1104–1107.

Prichard W. and Bentum I., 2009, *Taxation and Development in Ghana: Finance, Equity and Accountability*, Tax Justice Network Intl. Secretariat, Chesham.

Public Agenda, 2009, 'Oil will be hot potato for energy minister', *Public Agenda*. Available at: www.ghanaweb.com/public_agenda/article.php?ID=12727 (accessed 9 June 2009).

Public Agenda, 2010a, 'Ghana: Making oil a blessing through local participation', The Norwegian Council for Africa (Afrika.no), Africa News Update, 16 April.

Public Agenda, 2010b, 'Don't save oil money', *Public Agenda*, 8 October.

Public Agenda, 2010c, 'Citizens group oppose oil-backed loans', *Public Agenda*, 10 December.

Public Interest and Accountability Committee (PIAC), 2012, *Report on Petroleum Management for 2011*, PIAC, Accra.

Public Interest and Accountability Committee (PIAC), 2013, *Report on Petroleum Management for the period January 1 to June 30, 2012*, PIAC, Accra.

Pullen J., 2012, 'Henry George on property rights in land and land value: Equal and private, or common and public?' in Aspromourgos T. and Lodewijks J., eds, *History and Political: Essays in Honour of P.D. Groenewegen*, Routledge, London and New York, pp. 118–138.

Pullen J., 2013, 'An essay on distributive justice and the equal ownership of natural resources', *American Journal of Economics and Sociology*, vol. 75, no. 5, pp. 1044–1074.

Quainoo-Arthur R., 2009, 'The district assemblies common fund: Record and challenges', in ILO, ed., *Financing Local Economic Development: Old Problems, New Strategies*, ILO, Geneva, pp. 54–60.

Quarcoopome S.S., 1991, 'The politics and nationalism of A.W. Kojo Thompson: 1924–1944', *Research Reviews NS*, vol. 7, nos 1 and 2, pp. 11–21.

Randall A., 1975, 'Property rights and social microeconomics', *Natural Resources Journal*, vol. 15, October, pp. 729–747.

Rémy M., 1997, *Ghana Today*, 4th edn, Jaguar, Paris.

Reporters Without Borders, 2012, 'Press freedom index 2011–2012'. Available at: http://en.rsf.org/spip.php?page=classement&id_rubrique=1043 (accessed 22 May 2012).

Reporters Without Borders, various years, 'Press freedom index various years' Available at: http://en.rsf.org/press-freedom-index-2013,1054.html (accessed 4 July 2013).

Reynolds E., 1974, 'The rise and fall of an African merchant class on the Gold Coast 1830–1874', *Cahiers d'études africaines*, vol. 14, no. 54, pp. 253–264.

Reynolds E., 1975, 'Economic imperialism: The case of the Gold Coast', *Journal of Economic History*, vol. 35, no. 1, pp. 94–116.

Ribera-Fumaz R., 2009, 'From urban political economy to cultural political economy: Rethinking culture and economy in and beyond the urban', *Progress in Human Geography*, vol. 33, no. 4, pp. 447–465.

Rondinelli D., 1981, 'Government decentralisation in comparative perspective', *International Review of Administrative Sciences*, vol. 2, pp. 133–145.

Ross M., 2001, 'Does oil hinder democracy?' *World Politics*, vol. 53, no. 3, pp. 325–361.

Ross M., 2003, *Nigeria's Oil Sector and the Poor*, DFID, London.

Ross M., 2006, 'A closer look at oil, diamonds and civil war', *Annual Review of Political Science*, vol. 9, pp. 265–300.

Ross M., 2008, 'Oil, Islam and women', *American Political Science Review*, vol. 102, no. 1, pp. 107–123.

Rosser A., 2004, 'Why did Indonesia overcome the resource curse?' IDS Working Paper 222, Brighton.

Rosser A., 2006, 'Escaping the resource curse', *New Political Economy*, vol. 11, no. 4, pp. 557–570.

Rupp S., 2013, 'Ghana, China, and the politics of energy', *African Studies Review*, vol. 56, no. 1, pp. 103–130.

Sachs J. and Warner A., 1995, 'Natural resource abundance and economic growth', National Bureau of Economic Research Working Paper Series no. 5398, Cambridge MA.

Sachs J. and Warner A., 2001, 'The curse of natural resources', *European Economic Review*, vol. 45, pp. 827–838.

Saginor J. and McDonald J.R., 2009, 'Eminent domain: A review of the issues', *Journal of Real Estate Literature*, vol. 17, no. 1, pp. 3–43.

Sah S., 2008, 'Tsatsu's conviction stirs debate', *Daily Graphic*, 20 June.

Sala-i-Martin X. and Subramanian A., 2012, 'Addressing the natural resource curse: An illustration from Nigeria', *Journal of African Economies*, vol. 22, no. 4, pp. 570–615.

Salleh A., 2011, 'Making the choice between ecological modernisation and living well', *Journal of Australian Political Economy*, no. 66, summer, pp. 119–143.

Same A., 2009, 'Transforming natural resource wealth into sustained growth and poverty reduction: A conceptual framework for sub-Saharan African oil exporting countries', Policy Research Working Paper no. WPS4852, Washington DC.

Saminu Z.R., 2009, 'Oil experts advise prof. Mills Mr. President, watch out! Oil money is coming, but corruption will follow', *Chronicle*, vol. 18, no. 172. Available at: www.ghanaian-chronicle.com/thestory.asp?id=12280&title=Oil%20Experts%20advice%20Professor%20Mills (accessed 9 June 2009).

Saminu Z.R., 2011, 'Ghanaian businesses in Takoradi collapsing', *Ghanaian Chronicle*, 22 February.

Sandbakken C., 2006, 'The limits to democracy posed by oil rentier states: The cases of Algeria, Nigeria and Libya', *Democratization*, vol. 13, no. 1, pp. 135–152.

Sasraku F.M., 2013, 'Petroleum economics: Ghana's petroleum tax regime and its strategic implications', in Appiah-Adu K., ed., *Governance of the Petroleum Sector in an Emerging Developing Economy*, Gower Applied Research, Ashgate, Surrey and Burlington VT, pp. 163–174.

Schlater J., 1951, *Private Property: The History of an Idea*, Russell and Russell, New York.

Schoneveld G., German L., and Nutakor E., 2011, 'Land-based investments for rural development? A grounded analysis of the local impacts of biofuel feedstock plantations in Ghana', *Ecology and Society*, vol. 4, no. 16: 10, ISSN 1708–3087.

Segal P., 2011, 'Resource rents, redistribution, and halving global poverty: Resource dividend', *World Development*, vol. 39, no. 4, pp. 475–489.

Sekondi-Takoradi City Council, 1963, Compendium of articles and speeches on the elevation of Sekondi-Takoradi to 'city' status, Sekondi-Takoradi City Council, Sekondi.

Sekondi Takoradi Metropolitan Assembly (STMA), 2006, 'About this metropolis: Brief history'. Available at: http://stma.ghanadistricts.gov.gh (accessed 24 June 2012).

Sekondi Takoradi Metropolitan Assembly (STMA), 2010, *Draft Medium-Term Development Plan (2010–2013)*, STMA, Sekondi.

Sekondi Takoradi Metropolitan Assembly (STMA), 2011a, 'The composite budget of the Sekondi Takoradi Metropolitan Assembly for the 2012 fiscal year', STMA, Sekondi.

Sekondi-Takoradi Metropolitan Assembly (STMA), 2011b, 'Sekondi-Takoradi Metropolitan Area: Agricultural sector'. Available at: http://ghanadistricts.com/districts/?r=5&_=132&sa=2877 (accessed 25 July 2011).

Sekondi-Takoradi Metropolitan Assembly (STMA), 2012a, *Annual Report of the Physical Planning Department*, STMA, Sekondi.

Sekondi-Takoradi Metropolitan Assembly (STMA), 2012b, *Draft Structure Plan for Sekondi-Takoradi*, STMA, Sekondi.

Sekondi-Takoradi Metropolitan Assembly (STMA), 2013, 'The composite budget of the Sekondi Takoradi Metropolitan Assembly for the 2013 fiscal year', STMA, Sekondi.

Selby A.K., 1988, 'Review of "the rentier state"', *Bulletin of the School of Oriental and African Studies*, vol. 51, no. 3, pp. 555–556.

Servant J., 2003, 'Africa: External interest and internal insecurity: The new Gulf oil states', *Review of African Political Economy*, vol. 30, no. 95, pp. 139–142.

Shaxson N., 2007, 'Oil, corruption and the resource curse', *International Affairs*, vol. 83, no. 6, pp. 1123–1140.

Shaxson N., 2009, 'Angola's homegrown answers to the "resource curse"', in Lesourne J. and Ramsay W., eds, *Governance of Oil in Africa: Unfinished Business*, Ifri, Paris, pp. 51–102.

Silver J., 1978, 'Class struggles in Ghana's mining industry', *Review of African Political Economy*, no. 12, pp. 67–86.

Sittie R., 2006, 'Land title registration: The Ghanaian experience', Paper presented at the XXIII FIG Congress, Munich, Germany, 8–13 October.

Smith A., 1776, *The Wealth of Nations: An Inquiry into the Nature and Causes of the Wealth of Nations*, W. Strahan and T. Cadell, London.

Smith N., 2002, 'New globalism, new urbanism: Gentrification as global urban strategy', *Antipode*, vol. 34, no. 3, pp. 427–450.

Smith-Asante E., 2011, 'Ghana's first crude oil sold above $90 per barrel: Tullow', *Ghana Business News*, 11 January.

Smoke P., 2003, 'Decentralisation in Africa: Goals, dimensions, myths and challenges', *Public Administration and Development*, vol. 23, pp. 7–16.

Snapps O.J., ed., 2011, 'Africa-based oil exporting companies: The case of Nigeria', *American Review of Political Economy*, vol. 9, no. 2.

Solow M.R., 2008, 'The economics of resources or the resources of economics', *Natural Resources Policy Research*, vol. 1, no. 1, pp. 69–82.

Songsore J., 2011, *Regional Development in Ghana: The Theory and the Reality*, Woeli Publishing Services, Accra.

Spies-Butcher B., Paton J., and Cahill D., 2012, *Market Society: History, Theory, Practice*, Cambridge University Press, Melbourne.

Spivak G.C., 1999, *A Critique of Postcolonial Reason: Toward a History of the Vanishing Present*, Harvard University Press, Cambridge MA.

Staff Writers, 2008, 'Russian oil company invests US$100m in Ghana', *Afrol News*, 2 November. Available at: www.afrol.com/articles/29222 (accessed 2 November 2008).

Stilwell F., 1992a, *Understanding Cities and Regions: Spatial Political Economy*, Pluto Press, Sydney.

Stilwell F., 1992b, *Reshaping Australia; Urban Problems and Policies*, Pluto Press, Sydney.

Stilwell F., 2006, *Political Economy: The Contest of Economic Ideas*, 2nd edn, Oxford University Press, Melbourne.

Stilwell F., 2011, 'The condition of labour, capital and land', Paper presented at the Conference of the Association for Good Government', Sydney, Australia, 16 July.

Stilwell F., 2012a, *Political Economy: The Contest of Economic Ideas*, 3rd edn, Oxford University Press, Melbourne.

Stilwell F., 2012b, 'Marketising the environment', *Journal of Australian Political Economy*, no. 68, summer, pp. 108–127.

Stilwell F. and Jordan K., 2004, 'The political economy of land: Putting Henry George in his place', *Journal of Australian Political Economy*, no. 54, December, pp. 119–134.

Swing K., Davidov V., and Schwartz B., 2012, 'Oil development on traditional lands of indigenous peoples: Coinciding perceptions on two continents', *Journal of Developing Societies*, vol. 28, no. 2, pp. 257–280.

Taabazuing J., Luginaah I., Djietror G., and Otiso K.M., 2012, 'Mining, conflicts and livelihood struggles in a dysfunctional policy environment: The case of Wassa West District, Ghana', *African Geographical Review*, vol. 31, no. 1, pp. 33–49.

Tait D., 1951, 'Review of report on a social survey of Sekondi-Takoradi', *Man*, 51, December, p. 172.

Taper B., 1980, 'Charles Abrams in Ghana', *Habitat International*, vol. 5, nos 1 and 2, pp. 49–53.

Tawiah F., 2006, 'Tema and Takoradi Harbor: Good intentions, unintended consequences', 12 October. Available at: www.ghanaweb.com/GhanaHomePage/features/artikel.php?ID=111992 (accessed 6 October 2013).

Tibaijuka A.K., 2009, *Building Prosperity: Housing and Economic Development*, Earthscan, London.

Tiepolo M., 1996, 'Brazzaville', *Cities*, vol. 13, no. 2, pp. 117–124.

Tiesdell S. and Allmendinger P., 2003, 'Aberdeen', *Cities*, vol. 21, no. 2, pp. 167–179.

Tipple A.G., 1988, 'The history and practice of rent controls in Kumasi, Ghana', INURD WP # 88-1, 41088.

Tipple A.G., Korboe D., and Garrod G., 1997, 'Income and wealth in house ownership studies in urban Ghana', *Housing Studies*, vol. 12, no. 1, pp. 111–126.

Tipple G. and Speak S., 2009, *The Hidden Millions: Homelessness in Developing Countries*, Routledge, London.

Tjora A.H., 2006, 'Writing small discoveries: An exploration of fresh observers' observations', *Qualitative Research*, vol. 6, no. 4, pp. 429–451.

Transparency International, 2006, *Corruption Perception Index*, Transparency International Secretariat, Berlin.

Transparency International, 2007, *Corruption Perception Index*, Transparency International Secretariat, Berlin.

Transparency International, 2008, *Corruption Perception Index*, Transparency International Secretariat, Berlin.

Transparency International, 2009, *Corruption Perception Index*, Transparency International Secretariat, Berlin.

Transparency International, 2010, *Corruption Perception Index*, Transparency International Secretariat, Berlin.

Transparency International, 2011, *Corruption Perception Index*, Transparency International Secretariat, Berlin.

Transparency International, 2012, *Corruption Perception Index*, Transparency International Secretariat, Berlin.

Tsey C.E., 1986, 'Gold Coast Railways: The making of a colonial economy, 1879–1929', PhD thesis submitted to the Faculty of Social Sciences, University of Glasgow, Glasgow.

Tsey K., 2013, *From Head-Loading to the Iron Horse: Railway Building in Colonial Ghana and the Origins of Tropical Development*, Langaa Research and Publishing Common Initiative Group, Cameroon.

Tullock G., 1967, 'The welfare costs of tariffs, monopolies and theft', *Western Economic Journal*, vol. 5, pp. 224–232.

Tullow Ghana Ltd, 2009, *Golden Jubilee Field Phase 1 Development Draft Environmental Impact Statement*, Tullow Ghana Ltd, Accra.

Tullow Ghana Ltd, 2010a, *Annual Reports and Accounts*, Tullow Ghana Ltd, Accra.

Tullow Ghana Ltd, 2010b, 'Managing fishing intrusion in Ghana'. Available at: www.tullowoil.com/index.asp?pageid=363&category=ehs&year=2010&casestudyid=16 (accessed 18 March 2011).

Tullow Ghana Ltd, 2010c, *Corporate Social Responsibility Report: Creating Shared Prosperity*, Tullow Oil, Accra.

Tullow Oil, 2012, *Local Content Brochure: Supporting Local Business*, Tullow Oil, Accra.

Tuma E.H., 1971, *Economic History and the Social Sciences: Problems of Methodology*, University of California Press, Berkeley and Los Angeles CA.

Ubink J., 2007, 'Traditional authority revisited: Popular perceptions of chiefs and chieftaincy in peri-urban Kumasi, Ghana', *Journal of Legal Pluralism*, vol. 55, pp. 123–161.

Ubink J., 2008, 'Negotiated or negated? The rhetoric and reality of customary tenure in an Ashanti village in Ghana', *Africa*, vol. 78, no. 2, pp. 264–287.

Ugor P.U., 2013, 'Survival strategies and citizenship claims: Youth and the underground oil economy in post-amnesty Niger Delta', *Africa: The Journal of the International African Institute*, vol. 83, no. 2, pp. 270–292.

Ukaga O., Ukiwo U.O., and Ibaba I.S., 2012, 'Conclusion: Reclaiming politics and reforming governance: Options for sustainable peace and development in the Niger Delta', in Ukaga O., Ukiwo U.O., and Ibaba I.S., eds, *Natural Resources, Conflict, and Sustainable Development: Lessons from the Niger Delta*, Routledge, London, pp. 152–166.

UNDP, 2010, *Human Development Report 2010*, Palgrave Macmillan, New York.

UNDP, 2013, *Western Region Human Development Report*, UNDP, Accra.

UN-HABITAT, 2008a, *State of African Cities Report*, UN-HABITAT, Nairobi.

UN-HABITAT, 2008b, *State of the World's Cities 2008/2009*, Earthscan, London.

UN-HABITAT, 2010, *State of African Cities Report*, UN-HABITAT, Nairobi.

UN-HABITAT, 2011, *Ghana Housing Profile*, UN-HABITAT, Nairobi.

Van der Ploeg F. and Poelhekke S., 2007, 'Volatility, financial development and the natural resource curse', EUI Working Papers ECO 2007/36, Italy.

Vanwesenbeeck I., 2011, 'High roads and low roads in HIV/AIDS programming: Plea for a diversification of itinerary', *Critical Public Health*, vol. 21, no. 3, pp. 289–296.

Vearey J., 2013, 'A response to "migrant Nigerian sex workers and feminism"', *Writing Invisibility: Conversations on the Hidden City*, Mail and Guardian, Johannesburg, pp. 123–124.

Vidyasagar D., 2005, 'Global minutes: Oil purse or oil curse?' *Journal of Perinatology*, vol. 25, pp. 743–744.

Voices of Ghana, 2013, 'Britain pledges free port to service Ghana's burgeoning oil industry as Tullow project approved', 20 June.

Vokes R., 2012, 'The politics of oil in Uganda', *African Affairs*, vol. 111, no. 443, pp. 303–314.

Walsh A., 2012, 'After the rush: Living with uncertainty in a Malagasy mining town', *Africa: The Journal of the International African Institute*, vol. 82, no. 2, pp. 235–251.

Watt P., 2004, 'Financing local government', *Local Government Studies*, vol. 30, no. 4, pp. 609–623.

Watts M., 2004, 'Resource curse? Governmentality, oil and power in the Niger Delta, Nigeria', *Geopolitics*, vol. 9, no. 1, pp. 50–80.

Watts M., 2006, 'Empire of oil: Capitalist dispossession and the scramble for Africa', *Monthly Review*, September, pp. 1–17.

Wenar L., 2008, 'Property rights and the resource curse', *Philosophy and Public Affairs*, vol. 36, no. 1, pp. 2–32.

White H.P., 1955, 'Port developments in the Gold Coast', *Scottish Geographical Magazine*, vol. 71, no. 3, pp. 170–173.

Wilkinson K., Reynolds R., Thompson J., and Ostresh J., 1982, 'Local social disruption and Western energy development: A critical review', *Pacific Sociological Review*, vol. 25, no. 3, pp. 275–296.

Wolfinger N.H., 2006, 'On writing fieldnotes: Collection strategies and background expectancies', *Qualitative Research*, vol. 2, no. 1, pp. 85–93.

World Bank, 1989, *Sub-Saharan Africa: From Crisis to Sustainable Growth: A Long-Term Perspective Study*, World Bank, Washington DC.

World Bank, 1992, *Governance and Development*, World Bank, Washington DC.

World Bank, 1994, *World Development Report: Infrastructure for Development*, Oxford University Press, Oxford.

World Bank, 2003, *World Development Report 2004: Making Services Work for People*, Oxford University Press, Oxford.

World Bank, 2009, 'Economy-wide impact of oil discovery in Ghana', *World Bank Report*, No. 47321-G H.

World Bank, 2010, 'City of Accra, Ghana: Consultative citizens' report card', World Bank, Washington DC.

World Bank, 2013, 'CPIA Africa: Ghana'. Available at: http://datatopics.worldbank.org/cpia/country/ghana (accessed 4 July 2013).

Wright E., 2010, *Envisioning Real Utopias*, Verso, London.

Wutoh A.K., Kumoji E.K., Xue Z., Campusano G., Wutoh R.D., and Ofosu J.R., 2009, 'HIV knowledge and sexual risk behaviors of street children in Takoradi, Ghana', *AIDS and Behavior*, vol. 10, no. 2, pp. 209–215.

Xinhua, 2010, 'U.S. oil company refuses to pay fine for oil spill off Ghana', *Xinhua*, 21 September.

Yalley P.P. and Ofori-Darko J., 2012, 'The effects of Ghana's oil discovery on land and house prices on communities nearest to the oil filed in Ghana (case study: Kumasi and Sekondi-Takoradi)', West Africa Built Environment Research (WABER) Conference, Abuja, Nigeria, 24–26 July.

Yalley P.P.-K., Atanga C., Cobbinah J.F., and Kwaw P., 2012, 'The impact of the Ghana's oil discovery on land investment and its implications in the people, agriculture and the environment (case study: Cape Three Points, Ghana)', *Journal of Environmental Science and Engineering*, vol. B 1, pp. 922–930.

Yates D., 2012, *The Scramble for African Oil: Oppression, Corruption, and War for Control of Africa's Natural Resources*, Pluto, London.

Yeboah E. and Obeng-Odoom F., 2010, '"We are not the only ones to blame": District assemblies' perspectives on the state of planning in Ghana', *Commonwealth Journal of Local Governance*, no. 7, November, pp. 78–98.

Yeboah I.E.A., 2000, 'Structural adjustment and emerging urban form in Accra, Ghana', *Africa Today*, vol. 47, no. 2, pp. 61–89.

Yeboah I.E.A., 2003, 'Demographic and housing aspects of structural adjustment and emerging urban form in Accra, Ghana', *Africa Today*, vol. 50, no. 1, pp. 107–119.

Yeboah I.E.A., 2005, 'Housing the urban poor in twenty-first century Sub-Saharan Africa: Policy mismatch and a way forward for Ghana', *GeoJournal*, vol. 62, pp. 147–161.

Yeboah I.E.A., Codjoe S.N.A., and Maingi J.K., 2013, 'Emerging urban system demographic trends: Informing Ghana's national urban policy and lessons for Sub-Saharan Africa', *Africa Today*, vol. 60, no. 1, pp. 98–124.

Yeboah K., 2012, 'Experts call for greater transparency in oil industry', *Daily Graphic*, 21 December.

Yelpaala K., 2004, 'Mining, sustainable development, and health in Ghana: The Await case-study', Research conducted in Ghana from July to August 2003, and sponsored by the Brown University Luce Environmental Scholars Program. Available at: www.watsoninstitute.org/ge/watson_scholars/Mining.pdf (accessed 9 June 2009).

YoungHoon K., 2011, *New Town and Tourism: Development Planning for Ahanta West District*, KWAAK E.S.P.R.I., Seoul.

Index

Page numbers in *italics* denote tables.

Milton Keynes UK
Ingram Content Group UK Ltd.
UKHW031147141024
449569UK00024B/997

9 781138 626072